THE SECRET OF STONE

A new light on the Round Towers of Ireland

CHRISTOPHER FREELAND PH. D.

Self-published by Christopher Freeland, thaitoff@gmail.com

ISBN: 978-1-9161597-1-6

Acknowledgements

Where does one start? So much patience, kindness and sympathy from family, friends and acquaintances. People I scarcely know and others who have shared a common path for yonks.

THANK YOU ALL, but especially to Jerm!

*Dedicated to those industrious masons
who toiled on the round towers and
sacrificed their well-being for
the benefit of their fellows
and future generations.*

CONTENTS

FOREWORD

There have been myriad attempts made to grapple with the mystery of the round towers of Ireland over the many hundreds of years since their construction, and while some have come maddeningly close to making some kind of sense of them, their true purpose has eluded every mind that's ever been bold enough to scrutinise them. Until, perhaps, now.

Using the humble and profoundly accurate method of radiesthesia, ably assisted by his nimble intellect and tireless curiosity, Christopher Freeland has managed to pierce the enigma of the ancient towers and deliver to us a sound understanding of their raison d'être.

The official dogma that it was the church that built the towers is deftly dismantled in this single volume by Freeland's theory, which can not only be demonstrated, but actively experimented.

Standing on the shoulders of the ancient stone-masons, who had a firm understanding of the flow of energy in the earth, Freeland helps us come to grips with Nature's alchemy of stone, water and magnetism.

Magnetism is one of the fundamental forces at work on our earth. Even something as apparently inert as stone has a positive and negative polarity. When faced with the fact that every single stone in each round tower is laid with the negative magnetic polarity upwards, and furthermore, that all round towers in Ireland are located exactly above the crossing of underground waterways, the secret of this arcane alchemy begins to unmask itself.

In his erudite and elegant prose, Freeland brings us through many elucidating moments of his own discoveries and experiments with radiesthesia, frequency and magnetism, to deliver a persuasive and gratifying conclusion that will not only alter the way you look at the round towers of Ireland, but it will alter the way you look at nature itself.

More than a singular work, this little book is a polished gem of concentrated wisdom from a man whose findings have been collected on the empirical path by an open and curious mind.

Jeremy Massey,

West Cork, 2023.

INTRODUCTION

The development of any culture is directly related
to the understanding of its environment
– Callum Coates

If we neglect to investigate any aspect of what we <u>are</u> able to observe and comprehend, I believe that in all probability we are doomed to regurgitate what others have said, repeat their narrative and, sadly, never be any the wiser. In other words, an end to effective scientific research, which I would hasten to add in the case of the Round Towers of Ireland and Scotland merits a broader search for their purpose, especially as they have never been subject to anything more in the last few hundred years than sincere but forlorn guesswork, or at the worst, literary kow-towing to the imposed paradigm.

Wholistically, the subject of this essay is stone, but with a distinct emphasis on the application unique to the Round Towers, for these latter incorporate attributes which, I venture, are explored here for the first time in human history, my presumptuous belief knows no bounds!

Stone is the essence of the earth, along with water, and although much appreciated for its properties since time immemorial, it is not exactly the sort of thing that excites interest, let alone fires the imagination. So hopefully, by bringing to light a few realities on this superficially mundane substance, the idea is to share an insight, not fully appreciated in the collective conscious, of an inherent property of stone, specifically its magnetic capacity, its close relationship with water, and its gracious benefit. If we neglect – as we systematically do – such key facts, we are doomed to remain in nescience and miss out on some of the advantages that our ancestors were fully aware of, and generously bequeathed us.

This book is a small contribution to what deserves to be a full-blown investigation, if ever there is to be one, and is based on a few items of evidence that have, for one reason or another, escaped the attention of all other observers. These truths will not be to the taste of many, mainly because they fall totally outside of the mainstream story, secondly, consideration is given to some essential but neglected factors in the affairs of architecture, and thirdly, those factors by their very nature do not fall neatly into what is commonly accepted in the west as the scientific paradigm. What is more, the method used to determine these special characteristics is unorthodox, even described by some of its detractors as pseudo-scientific. Having said that, I would ask you at least to hear me out and then after careful consideration, please draw your own conclusions, and even better, start your own inquiry into the evidence we have at our disposal so as to understand what possibly happened to this science – for that is what it must have amounted to for a fairly lengthy period in human history – to have been eclipsed. Finally, for the curious and the practical-minded reader, a look at the possible applications in our modern age.

The facts presented here are up until this present moment in time, I believe, universally ignored for the most part, except some perchance among water-divining circles, therefore may require a period of reflection and experimentation before gaining currency, or rather establishing a renewal with ancient tradition.

There is no specific intention to disprove other points of view, although that will invariably be the result, due to illogical or the total lack of reasoning, expressed by the protagonists themselves in their haste to reiterate the accepted paradigm and garner further acceptance from the herd. However, as the legal maxim would have it: "No one is bound to give information about things he is ignorant of, but every one is bound to know that which he gives information about."

Every effort has been taken to be as all-inclusive as possible of the data and literature concerning the Round Towers, so please excuse my shortcomings, and what might be seen as a contrarian attitude, but a wholistic method of inquiry cannot exclude observation of what is actually present, irrespective of whether such meets mainstream approval or not. There is no desire to impose a belief, for we are highly unlikely to know with precision what was in the mind of the beings that built these singular stone structures which are still with us after many hundreds of years, in fine fettle but unfortunately not in prime working order. One thing is for sure, if we blindly repeat what others say and close our minds to relevant facts, intolerance and prejudice will be close at hand.

What follows is a shot at quite what the significance of these facts might be.

CF, Kilcrohane, February 2023.

CHAPTER 1
THE USE OF STONE BY HUMANS

Even the most insensitive soul cannot fail to register some form of sentiment when entering a Gothic cathedral, or in the presence of a megalith, especially an array of stones such as Carnac in Brittany, Stonehenge in the UK, or Loughcrew in Ireland.

But what do we know about stone? What is it that resonates within us when in the presence of a stone circle? Is there something which we sense, but lacking an understanding of that specific sense which somehow enables the connection, and due to a complete lack of education in this respect, causes us to relegate the experience to one more, but rapidly overlooked gut-feeling?

Despite the extensive use of stone throughout the world in just about every culture, there is little or no information available allowing for an understanding of its properties, let alone how those characteristics can, or have been applied. This book started as an essay, an attempt to clarify what I had been researching and experimenting with for a number of years, until a chance discovery when examining one of the Irish round towers. That then led to a winding but essential journey, along the majestic paths of Pelasgian/Pelasgic or Cyclopian architecture, civilisations long past with their occulted yet intimated techniques, as well as a few dead-ends which had to be explored so as to exorcise certain possibilities.

All due, however, to an apparently banal and simple insight regarding stone.

While the thrust of what I am proposing here is multiple and will hopefully be dealt with in detail in due course, most of what I say is based on simple observation, historical and scientific research and experiment, with the addition of an inevitable element of intuition as we are dealing with (publicly) undocumented, obscure (because obscured perhaps) topics, such as:

1. The ability and use of stone to generate a feel-good sensation.
2. The "magnetic" field of force present in stone.
3. Antennae have a specific length in relation to their function.
4. Geographic positioning of stones over geological phenomena have differing effects.
5. Pattern or layout enables resonance.
6. The interaction between cosmic and telluric forces can be focalized.
7. Energy can be beneficially modified by using the characteristics of specific locations.
8. Energy can be beneficially modified by using specific materials.
9. Shape and pattern can and does alter the flow of energy's movement.
10. The benefits to health, ours and that of the environment.
11. The means to achieve those benefits.

An intriguing trinity which may appear idiosyncratic but will be developed further

12. The Geographic principle.
13. The Polarity principle.
14. The Cross principle.

Stone is the origin of water in some traditions, the bearer of memory due to its intimate relationship with water as the aeons pass; for others, the partner to water, sharing the exemplary humility some strive for; elsewhere, towering above us all, stone is the paragon of excellence. All of these combined, and more.

It is not to everyone's taste no doubt to experiment with stone, and while one may even be incapable of recognising this capacity of stone to create the 'feeling' mentioned above, that makes absolutely no difference to that ability. The Japanese and Chinese (prior to 1949) landscape gardeners, and the former still do, used the properties of diamagnetism and paramagnetism to generate an enhanced environment for plant growth, but not just for the benefit of the plants. That sense of well-being knocks on to humans and animals due to the interconnection or sharing of the vital force we all benefit from. I would propose that there is a very strong chance that is what the ancients were deliberately doing when constructing the monuments of stone that we still have amongst us today.

Unfortunately or so it seems, either there has been a constant and intense effort throughout the ages to occult the importance of the magnetic field for one and all of the earth's flora and fauna. Or, as Abraham Liboff said in 2013: "The interaction of weak magnetic fields, with intensities on the order of the geomagnetic field, is a very interesting subject that only recently, in the last few decades, has received much scientific attention." I would put my money on ignorance and neglect, it is just too obvious that stone has magnetic properties, but if you do not know how to detect, then deploy them, they fall into disuse and the trashcan of history.

Without instruments to measure the low magnetic fields of force it is nigh on impossible to prove to the satisfaction of one and all, that a field of force impacts the individual's nervous system, even though such a sensation might well be felt physically and the beneficial effects experienced. Such measuring devices are not commonly available apart from some very expensive laboratory equipment which is not readily portable. Having said that, it is a well-known phenomenon among alternative therapists that magnetism as well as the laying on of hands of someone with a strong magnetic field induces not only relaxation, but sometimes a cure. A sense of peace and calm. Under any regime, at any period of history, that is a much-prized quality valued by any society with a focus on harmony and well-being.

Even if we do not understand how such resultant well-being is accomplished, and it is not just a faculty specific to a man or woman, it is commonly found to be a much appreciated facet of any site, building or location that generates it. In most cultures such spots become places of pilgrimage, power spots or sacred sites and their domination can become cause for struggle of the control of the surrounding population, or at least their attitude towards the source of that influence. There are numerous examples of this found around the world, especially with water sources which have healing or therapeutic properties. As we shall see further on, I think that is very much what happened when the church claimed authorship of the round towers, but as is well known in legal circles – the burden of proof lies upon him who affirms, not on him who denies. The only difference being that it is the supporters of the church that claim authorship whereas I merely request the proof – to no avail.

It appears there are governing principles in force here on this planet, and irrespective of what man thinks – and does, there is a very strong probability that this has always been so. Nature rules!

You might agree that one of the most impressive effects on entering a cathedral built during the medieval period, from 1170-1340 or so, is the sensation that is almost physical in its intensity. That sensation is not found on entering any other ecclesiastic building, nor in the same manner, whereas it _is_ encountered when crossing the threshold of certain ancient monuments that share some of the principles discussed here. Once again we are very much in the dark as to a reason, if any, for this.

As Philip Ball points out in his book *Universe of Stone,* the Gothic cathedral for Victor Hugo "was a social construction, a temple made for and by the people rather than decreed by an ecclesiastical elite."

There are few such pilgrimage sites in western Europe that are not dominated, controlled or claimed by the church, yet the we never question that authority. While it is possible that could be the case, would it be too much to have an explanation, some evidence?

Literature is indeed the most practical, when we know what we are reading, but time is not tender on papyrus, paper and leaves, so limits are soon reached, and we revert to the stones to gain insight. When we are fortunate enough to have them available, and they have not been removed, destroyed, buried, or used for other purposes, we have an open book before us. But what is the message, can we even read the script?

It is largely thanks to archaeological remains that we have a vague idea as to what our forebears were up to, as there seems to be little in the way of explanation from literature or mythology in any of the European traditions. In actual fact, there is nothing more than conjecture for saying that, and it is thanks to people like Alexander Thom (1894-1985), the Scottish researcher who did so much to demonstrate that the ancients were not only highly competent engineers and technicians, but he made a space for archeoastronomy, so proving that the standing stones serve a very distinct purpose, which regrettably is not accepted as gospel truth by mainstream science. As a result any ideas coming out of alternative archaeology can only be considered as hypotheses, which is what they have always been, and what is written here is no different save that it is confirmed and actively used in the Chinese and Indian traditions of *feng shui* and *vastu.* Accident, coincidence, synchronicity, design... are some of the expressions we employ to explain what we do not – and never will – understand in life. These pages are no exception as they recount a story of revelation and discovery (I trust not factual banality) of an unusual property of stone.

This hopefully makes for a more convincing argument backed as it is by practical applications which can be easily experimented, verified and applied, and hopefully contributes a dimension that is rarely if ever considered, but as a consequence, is simply rejected. There is no end to the scepticism born of modern 'science', and regrettably, no space granted to the sincere wholistic approach requiring an open mind which at the worst could provide a little insight.

It would seem that in ancient times, the potential of certain of these principles was better understood than today, and practical use was made of them. It goes without saying that humans have a streak of ingenuity, and from our vantage point of the present moment we look back at various stages in the past where people have used that inventiveness to incorporate the forces of Nature into their structures and practices in much the same way that is done today, well more or less. The difference being that we now know that interfering with planetary frequencies is fraught with consequence. There will always be those who, with the benefit of hindsight, or due to an innate common sense and decency, keep within the bounds of reason and maintain a healthy respect for Nature. By the same token, there will also be those determined to profit at any cost.

I can only echo Alexander Thom as he said in *Megalithic Lunar Observatories*, "We do not know the extent of Megalithic man's knowledge of geometry and astronomy. Perhaps, we never shall. He was a competent engineer. Witness how he could set out large projects to an accuracy approaching 1 in 1000, and how he could transport and erect blocks of stone weighing up to 50 tons."

The focus in this essay is on the "lower end" of the monumental scale, which includes the vast majority of stone, starting with the more common standing stones, obelisks, circles, menhirs, round towers, rather than the gigantic structures such as the pyramids found in Egypt, North and South America, China, Java, Bosnia, the Bermuda Triangle, Greenland, and so on.

Archaeologists, like every other -ist generated by modern science, are often obliged to restrict their investigations to the academically correct and acceptable criteria imposed by current dogma, and frequently by tenureship. Few academics dare venture outside that corral. There are some amateurs who have done and do so, but rarely do we hear of them in the mainstream, if at all. Obviously, that is a great shame because so many brilliant and curious minds are stymied from casting their insights into the public arena, and the fictional narrative is maintained. As in modern-day society, there is no longer room for an arena where open discussion can be held of ideas and discoveries that might be of benefit for the world at large.

In much the same way that no sensible Egyptologist – wishing to retain their livelihood – would mention that no bodies have ever been found in a pyramid, no orthodox researcher into the Irish Round Towers would dream of not mentioning a church in connection with the towers.

As a wholistic researcher, specialising solely in extracting a version of the truth from the evidence, using means both orthodox and unorthodox, I am free to roam at will outside the box. Any challenge to what is written here is most welcome so long as it is constructive (non-derogatory), enabling me and others to better understand, and conducive to greater precision in the subject matter.

At the end of the day, the most apparent feature of ancient human history is to be found via the use of stone. Throughout the known world, stone vestiges are to be found and reveal all sorts of information as to what life might have been like years ago. However, there is far more left concealed from our understanding, and if we persist in neglecting the factors discussed here, we are doomed to remain in the dark.

CHAPTER 2
THE RESEARCH METHOD

True Science deals with quality, or essence. You know when you are in your essence of being as a man or woman, for wherever your gaze falls you perceive perfection and it is harmonious, it is one. Lose that fundamental human value of grace with its intimate conviction that everything is in its just place, and a space is created which soon fills with quantity, the world of mental spawn. The quality of the one is seemingly overwhelmed by the quantity of multiplicity, and a struggle for harmony begins.

QUALITY INFORMATION

You might agree that information to be considered as worthwhile and useful must be practical, essential and authentic. Worthwhile because it benefits one and all, serving specific objectives; practical because it is accessible and simple of application; authentic because it is the truth. A possible definition of this expression would be any knowledge that reaches our awareness, thereby encouraging and reinforcing a sense of unicity, a practical, comprehensive gnosis which remains constant within the bounds of natural behaviour and events.

The ability of humans to access information of quality has always existed in divination. The acid test is naturally one of value – we are human after all, and of course values change over time, as a function of a host of factors. The most powerful influence over human values would seem to be emotional security, in other words feeling safe. Our instincts operate in that domain, striving for food, shelter, reproduction, sleep and security all providing for a sentiment of well-being once achieved. That has been the case throughout history, and one only has to look at the controlling interest in recent periods to appreciate that. For example, in our current phase of mass indoctrination that there is some horrific bug that is going to get us unless we stay at home or agree to a so-called vaccination; or just over a hundred years ago, we were fooled into massacring each other in the most barbaric event, that was repeated twenty five years later. Without human acquiescence on a massive scale such incidents cannot happen. They repeatedly occur because we allow ourselves to be emotionally manipulated, the rest is then easy for the puppet-masters to coordinate, backed by the powerful resources of societal pressure, influenced by the media and financial stimulus. No matter who the masters are, I am simply pointing out that this is a historic pattern based on contrived information. However, that is not what one would qualify as quality information.

That remark of course raises the question of what is essential in the way of knowledge which could make it information that we would deem of quality. Notions as to what that could or should be are probably as numerous as the minds posing the question. If one takes Nature as the guide, for she is the ruler, then what is harmonious wins the day, closely followed by what is useful and practical, so long as the aim is one of comprehensive, all-embracing harmony. We moderns have so much information at our disposal, and increasingly so of the superfluous variety, it is quite hard to discriminate what is good and bad for us. Particularly when 'good' and 'bad' are no longer subjects for us to decide upon, when the practical objectives have been turned to commercial gain by means of manipulation, then fed back to us to consume. The danger is that we throw the baby out with the water as we either focus on the objects of desire and greed, or simply give up to the overweening manipulation, and stop thinking. The latter is very much what could be taken as the cause of chaos, for chaos is a totally subjective affair.

Back to our sources of information, which deserve a closer look to see if we cannot surprise an insight as to how we actually acquire knowledge. If life is movement (as opposed to stasis), and movement is a function of the various forms of energy coming in from the cosmos, and the resultant interaction here on the surface of the earth, it would not be foolish to consider observation of these movements as a source of understanding, IF we are able to associate an event with a result. Regrettably, over the course of time, this ability to observe and synthesize the whole became a very binary affair, and we ended up with dogmatic cause and effect, the most basic misconception imaginable because lacking the subtle exchange offered by the wholistic approach, where everything has its place, its cycle of life and demise.

There are two very ancient methods of accessing information and both can be freely obtained with effort. As is so often the case with something that is for free, we look askance and tend to diminish the value, even disdaining and rejecting the offer. Yet, Nature provides us freely with life. It would be well to recall.

If you have not already guessed, these two methods are Radiesthesia and Astrology. They have been practised both openly and occultly, since time immemorial, in all civilisations and cultures. They are what I call sci-art – scientific arts – par excellence, the only ways to discover instantly what is going on and what is about to happen. In the hands of the talented practitioner, there are no secrets.

Astrology is the royal path, not because it is better in any way or form, but because the elite down through the ages have encouraged and used the services of astrologers with the necessary reputation and ability. There have always been astrologers working in close proximity to the crowned heads, as indeed nowadays, although not such a well-known fact, in every head office of the banking, insurance, brokerage, media, music, publishing industries and others too. Risk is not for those who can afford it, and it is generally a costly service.

Dowsers (radiesthesists) are systematically employed by mining, oil and gas firms.

Immense, but perhaps deliberate damage has been done to astrology. It has been vulgarized almost beyond recognition in the last one hundred-odd years. As is easily imaginable, the transmission of the immense detail involved in astrology over the ages, and from one culture to another in different languages and contexts resulted in variants and inevitable misinterpretations creeping in. The ancient practices of the Babylonians made their way down through the ages largely thanks to the Arabs, squeezing through the mesh of church and political strife until the seventeenth century when there was a revival in Europe, soon eclipsed by Newtonian 'science'. In 1914, however the Theosophist, Alan Leo (William Allan) instead of sticking to the already uncertain concepts, obscured astrological matters even more by inserting the egocentric psychological phantasms recently made fashionable by a few Austrians especially into his popular books, and the public was encouraged to accept the new-age nonsense. The objective, if that was really the case, was achieved and the ersatz version was successfully separated and presented as we have it today, while the real astrology continues in the hands of the few, even though totally accessible to those who spare the energy and time. Incidentally, the main reason for Leo doing this was to avoid doing time for witch-craft!

So the damage carries over to the unsuspecting public, now potentially deprived of a true sci-art unless a conscious search is made for a traditional astrologer who steers close to the essential, and avoids the psycho-babble.

RADIESTHESIA

The other form of divination, not as familiar as the planetary art, is Radiesthesia. This is the down-to-earth approach, as opposed to the more general and symbolic approach of astrology, requiring a questioning process provided by the human rational function, and the fundamental key which provides the answers, the all-aware, omnipresent Consciousness which is shared by all that exists, in every realm, at all times. In our anthropocentric way, or for convenience's sake, that is located in the heart, where the spark of life is to be found in the mammal.

It is the only means of accessing truth, or quality information, instantaneously, without any need to cast a chart, consulting of the ephemeris or computer.

Radiesthesia, despite this name having been invented in 1920 only, has been around for ever, no matter the name as we saw above. It involves a tedious training involving practice and yet more practice, a special vacuous mindset where you 'know' nothing, so it does not enjoy the same popularity as the astrology, nor the same renown.

The accomplished radiesthesist is able to provide an immediate answer to a question. This is not quite the case with astrology. While horary is a most useful form of astrology, it is definitely a task for the experienced practitioner, and it is highly unlikely that you would be prepared to pay 200 USD or so to find out the answer to a simple question of a humdrum nature, but it takes more time to prepare a chart and interpret the symbolism so is probably justified.

Radiesthesia is the 'intrinsic' method, definitely more ignored than the planetary art, yet sufficiently simple to be accessible to one and all.

When I say an intrinsic approach, this is in opposition to the input from the planets and cosmos. The frequential energy pouring into our environment makes for a constantly changing mix which is in all probability, quite impossible for an untrained human to fathom. But once the trained astrologer can appreciate the patterns, and make the correlation between cause and effect, recognising the interaction at work among the multiple factors involved, the picture becomes clearer, and we are 'in the know'.

It is somewhat different with radiesthesia, with a need to assess the context and to ask the right questions, a highly demanding task calling on a careful combination of intellectual ability and intuition, with a large dose of humility.

Both methods have encountered mixed fortunes as the ages have progressed, but ever since the age of enlightenment 'dawned' in the 18th century, the eclipse of both has apparently been nigh on total. Radiesthesia made a brief comeback between 1910 and 1950, especially in France. It was extensively used by French doctors in the early 50s as a very simple but efficient means to check if diagnosis and prescription were correct. That did not last long, but at least they were no longer burnt at the stake.

As Christopher Hitchens said: "What can be asserted without evidence cannot be dismissed without evidence." It is neither an easy task, nor without risk of incurring enmity and wrath, to upset the accepted paradigm.

We have now become so used to being told what is good for us, even the most intelligent people accept without a murmur as soon as a 'name' says it is good or bad, or a household brand is determined to be just what you need. A clear sign that our discrimination has been demolished.

Without the ability to discriminate, the individual cannot determine what is real, what is beneficial, and relinquishes their individual humanity to become a member of the gormless herd. This is not a matter of conspiracy, it is a simple observation of what a human can achieve in the absence of a moral compass.

It might be necessary here to develop the idea of Consciousness, which of course can only be intimated – it is what we all are. A useful manner of accessing the notion is by means of one of our senses. I would redefine these along the lines of the traditional Egyptian teaching where the human has 365 of them. I don't know where Aristotle lost the other 360 after spending about twenty years at school in Egypt, unless the plan was to reposition the human in a purely physical setting, and that worked very well.

However, in our domain of multiple senses, Unicity, could be said to be the foremost human sense, experienced by some as the feeling of being-at-one, atonement if you like. By nature, it is a continuity; has there ever been any doubt in your mind whether you exist? The doubt is one of form, not foundation, which gives rise to duality – me and you, and her offspring, multiplicity – the world. That sensation of oneness can never be affected. In none of the three states humans experience, viz. waking, dreaming and deep sleep, is that gnosis absent. But we tend to pass it by because it is so immanent, we do not affirm its existence, and soon become complacent due to the familiarity. Most of our socio-philosophical systems teach us to look outside for some form of governing influence, rather than applying introspection which would soon teach us that there is no inside, nor outside. Its most common name is Love. It is much easier to love your neighbour, as we are told is the right thing to do, when you know you ARE your neighbour!

That being-at-one sensation is, I believe, the result of feeling that you are in exactly the right place, which it would be reasonable enough to assume, happens when your physical body is in harmony with the surroundings. To do that, the correct balance of positive/negative, attraction/repulsion, opposites must presumably be in sync. In other words, the 'magnetic' component is as close to harmonious as possible. We will take a closer look at this 'magnetic' component shortly.

Human evolution, which of course exists for us anthropocentric species, depends totally on incoming cosmic frequencies – whether we like it or not, if perchance we can recognise that truth. Unfortunately for a little over a century and a half, we have managed to interfere with those frequencies, in the form of electricity and, more recently, electro-magnetic radiation, in such a way that certain categories of dis-ease have been introduced into our environment, but of course it is not our fault, so we do not even deign to investigate that possibility; although I suspect vested industrial interests ensure that will not happen in the near future.

Naturally enough, those frequencies accumulate in all elements, thanks to Nature's alchemy which aligns the component molecules over time, as a function of the type of element and surrounding conditions, resulting in an aggregate magnetic charge which varies from one type of substance to another depending on the specific locality, different periods of day and night, and all sorts of other factors, the finality is a weakening of the magnetic field, the earth and its denizens.

But what exactly do we know about how this energy-exchange really works? Rather than start a polemic which is unlikely to be resolved, can we assume a wholistic approach, an overall, non-differentiated, interconnected and communicating view, as much as that is feasible for a simple human to achieve?

And I think we have missed something. The problem in not recognizing the unicity of energy and its presence in all aspects, physical/subtle, kinetic/potential, lies in the non-observance of the underlying harmony that is to be found in all components that make up Nature. Observation is a forgotten, almost extinct art. We know of train-spotters, bird-watchers, astronomers, philatelists and a host of wonderful hobbyists whose leisure time is absorbed by their singular passion. There are not many Daoists in the neighbourhood whose lives are spent shadowing Nature!

One thing is certain: the ability to recognize the fact that we are an integral part of Nature, belonging to this phenomenal universe as components rather than as its masters, is rare, but essential if we hope to surprise her secrets.

There is a dearth of information concerning this wholistic aspect of energy in our western cultures, but that is not the case for other older but still current cultures. Such a state of affairs is deliberate on the part of those who govern us, otherwise there would be no reason for the restrictions as there are, for example with regard to frequency medicine, practised almost uniquely in China, and locally at that, because elsewhere it is 'illegal'. Although that does not help us progress in our quest, we must be aware of that situation and even more so, that such a restriction has been imposed on the common mortal since very early times, in one form or another.

If energy, as expressed in the living organism or the body, is understood as the vital force organizing the flow of blood and fluids, breathing, the nervous system and the vital organs, all operating thanks to a constant flow, we not only find affinity with the theory of the meridians in Traditional Chinese Medicine (TCM), or the three *dosha* (humours of wind, fire and mucus) in Ayurvedic medicine, but there is strong evidence that the ancient stone-masons had a firm understanding of the flow of energy in the earth. The common denominator of all the above is movement, no matter the speed.

Energy can only act and interact when movement occurs. Death is the absence of movement.

Who says movement says current, and a current of movement is said to generate what we vaguely call a 'magnetic field' in the immediate vicinity.

The modern paradigm functions on the principle of differentiation, as when determining the separation of potential from the kinetic, introducing six basic families of energy – thermal, chemical, electrical, and so on. What is very encouraging today is that there are even some schools of modern physics coming round to the idea of an interconnectedness of these different forms, either as a coherent totality or from the position of individual consciousness. This is wonderful but somewhat like reinventing the wheel for those familiar with thought systems from older civilizations; however, better late than never.

This is probably because the emphasis is constantly on the physically apparent manifestation with little or no attention given to the subtle processes that go on invisibly in those parts where we do not currently look.

For example, in 'magnetism'.

CHAPTER 3
'MAGNETISM' ITS FUNCTION & INHERENT PROPERTIES

A pragmatic definition of "magnetism" would be useful to guide us, but there is none to be found. It is as if the subject is booted into touch, to be dealt with at some later stage, another day. The orthodox scientific reference Britannica.com attempts an explanation with a remarkably lame: "All matter exhibits magnetic properties to some degree." That is then rather idiotically repeated on the net. Presumably this refers to the properties of attraction and repulsion. There is, however, no question of "where does it come from?", nor "how did it get there in the first place?" Instead, we are fobbed off with suppositions, theories and mind-boggling data, whereby monopoles, dipoles, electrons and atoms torque and spin through a bewildering array of the human-denominated categories of physics, biology, chemistry, geology, anatomy, and other branches of 'science', plus others for which we have not yet found names, yet all carefully measured in tesla or gauss. Welcome to the world of quantity!

In other words, as is so often the case in the current era, instead of assuming a potentially wholistic, open-minded approach with a broad consideration of inter-relationship with life in general, we isolate and deal with a single function, so that it can more readily be expressed as a quantity, but in so doing, we basically kill the event, in much the same way when a cell is placed under a microscope removed from its natural environment. Of course, it is an excellent idea having discovered a facet of a phenomenon to research and develop it within the limits of one's capacity. But to curtail further study is suspect, and it is hard to imagine a lamer statement than, "all matter exhibits magnetic properties to some degree." One cannot help but wonder about energy-directed weaponry and the applications of electromagnetic frequencies, which rely on a significant "magnetic" component in order to function.

Once again, we are obliged to beaver away behind the scenes in order to find ways to restore a little harmony in our environment, because it is clearly very damaging how neglect of the need to maintain a 'magnetic' balance has now become for Nature at large, humans, animals, plants and the atmosphere specifically.

Deliberately above "magnetism" has been placed between inverted commas. This is too vague an epithet for what we are trying to discover, so another term needs to be found having defined the various characteristics which make up this complex.

We have a tendency as humans to confuse the properties of an object with its functions. In doing that, however, we lose sight of the inherent characteristics as we become absorbed, as it were, by the power of the thing in question. Especially when that thing exhibits a capacity we are interested in, to the extent that we are able to ignore the disadvantages, especially when profit is involved in the application. In this day and age, the perfect example of this would be the mobile phone.

By the same analogy, what we know as magnetism has become something so banal that we have ceased all research into the how and wherefore, and fall for the lame excuse "we don't know exactly what it is but we know that it exists".

ORIGINS OF MAGNETISM

The thrust of this section is a proposal that the movement of energy flowing in from the cosmos, its accumulated effect on the earth, and subsequent flow through the earth's strata and components, are the probable cause of this phenomenon we call 'magnetism'. Such a theory would even support the amazing statement above, but for reasons that are regrettably apparent they are beyond the scope of this essay.

What could be more natural for a force to be developed when two sources of light – one very powerful 'electromagnetic' source, the other with a yet-to-be-explained presumably magnetic capacity, fluctuating as a function of their constituent components – have been revolving around our heads for several billion years? When you know that an electric current can be generated by turning a magnet wrapped in a coil of wire round and round, a little perspective is gained, or should be. The earth is/has become a magnet, the source of a power concealed in its entirety because we refuse to attribute a recognized, commonplace phenomenon to our own home. The circular movement performed by the two luminaries, day after day, has not been examined in this dynamo light (so to speak). There is a constant field of force evolving, unseen, unheralded but in plain sight.

Which came first, the magnetic material or the revolving force? I don't know, and do not care to know. What is important, I believe in this context, is the ultra simple fact that stone, which makes up the majority of our land mass as well as the firm bit below the seas, is affected to varying degrees by this force, as witnessed by what we know as the magnetic properties of attraction and repulsion. Of course, those two qualities are not unique to the attractor or the repeller, they are shared and interact as a function of their position one to the other, with other factors most probably playing their part, affecting the intensity. We will look at this characteristic concerning the degree of strength or otherwise, the properties and uses further on.

NATURE OF MAGNETISM

When one thinks about it, there is every reason to believe that magnetism could well be the primary force at work with regard to the development, maintenance and potential destruction of our environment. The constant flux of cosmic vibrations from the universe exchanging with the various components of our atmosphere should not be discounted as a cause for the charging of the contents with magnetism. The constant flow of billions of years of cosmic frequencies must have had some effect, and presumably a lasting one at that. But as with all tides, there is an ebb and flow.

According to received (rather imposed) science, the making of a magnet, or any body showing a magnetic force, involves the aligning of the atoms of whatever the material is. This, we are told, comes about due to the spin of the atoms as they are impacted by the flow of the current to which they are sensitive. The electrons all spin in one direction, with those at one pole spinning to the right, and the electrons at the other pole, spinning to the left. This apparently depends on the number of atoms that can be magnetized. In a stone, this is manifested by the dia- or paramagnetic strength, as a function of the native material. Although the electron spin of the atoms is aligned in a specific direction, remember that the spin from the pole at the other end will be in the opposite direction. Is it just a coincidence that the sun always rotates in the same direction?

The Egyptians, according to Manetho (3rd century BC), had a name for the power that presided over this circulating potency of the sun – the bone of Horus. The Greek Euripides, according to Plato, gave it the name *magnet*.

There is no mention of an inherent form of magnetism in this formal version of science shared in academia. The method of quantitative measurement or the gauss measurement of magnetic force is a function of the number of molecules in the material whose electrons can be magnetized or polarized. Absolutely no attention is given to any of the other properties of magnetism beyond the measured scale, they are not only sublimely ignored but the other factors are not given any space at all in the equation.

The physics of such a possibility will have to wait for another day, for it is neither within my ability to explain, nor do I have the infrastructure to prove such an idea with experimental evidence. Nevertheless, the fact remains for mainstream science and alternative technology that 'magnetism' is one of the fundamental forces at work on our earth, and its patterns and fashion of functioning deserve greater attention if we hope to surprise the truth.

So many aspects of this magnetic force thrust a logic upon us, which is ignored by the majority, but when quietly adopted and assimilated, provides huge advantages. This is where we can acquire benefit and use the presence of magnetism in our daily existence. What happens when magnetic cohesion declines? Things no longer stick together as they did, and gradually/instantly fall apart. Decomposition, and a return to source? The instances of quality, as I would call them, can be easily adopted to make life a joy rather than a bane: such as when accumulating a magnetic charge which can be so easily performed by a human with any form of magnetic recharging exercise, but especially *tai chi* or *qi gung* – designed to that purpose, resulting in reinforced vigour; or, the exchange of magnetism, as in the laying on of hands, when a transfer takes place, with the often surprising therapeutic results known to those who take part in such alternative practices; as also when grounding by walking barefoot, or using a grounding mat beneath the body during sleep or a brief period of rest.

One of the main points in this book is the essential importance of "magnetism". Essential because it is one of the fundamental components in the functioning of all that lives, and even present in what we consider as dead. It would not be a lie to say that what we term magnetism is vastly underestimated, for a resoundingly simple reason, its cause and prevalence in everything that exists is either neglected, ignored or, at worst, has been expressly occulted. That might appear to be a very provocative statement I know, but when one considers that basically no research is conducted – publicly at least – on magnetism, then there is justification to ask why, given that it is such a commonplace, yet seriously misunderstood phenomenon.

While magnetism is expressed in the human, it is nevertheless a very low field. as a result of the flow of blood, nerve and vital energy. Modern medicine measures the electric current in the most basic of diagnoses with the ECG or EEG. , and Tokyo Denki University in 1992 published their findings and the abstract of their paper states: This article describes the present status of research on biomagnetism, an interdisciplinary field of research involving biology, engineering, medicine, physics, psychology and other areas. Biomagnetic fields are caused either by electric currents in conducting body tissues such as the heart, the brain and muscles, or by magnetized material in lung contamination. These magnetic fields, although measurable, are so extremely weak that a superconducting quantum interference device (SQUID) magnetometer with ultra-high sensitivity is needed to detect magnetic flux generated outside the human body. This paper mainly focuses on the remarkable progress of research on biomagnetism involving magnetoencephalograms (MEGs) and magnetocardiograms (MCGs), and on the introduction of SQUID systems for measurement of biomagnetic fields.

On 26th December 2009, a paper published by the PTB Institute in Berlin, in conjunction with the American NIST, using an optical magnetic field sensor substantially different from the SQUID, found that it was suitable to measure biomagnetism in the picotesla range. Well, at least science now has a basis for some kind of standard.

Without going down too many rabbit holes, and to keep on the scent of stone, it might be timely to mention an anecdote of the use of stone as a therapeutic application.

Some years ago in Chiang Mai, I was called in to help a bed-ridden patient, very much a last resort type of thing in a desperate appeal for some small relief. Not knowing anything about the person or their condition, but much concerned when confronted with such total weakness, the situation screamed a need for minimum energy. Having recently discovered the "magnetic" properties of stone, I asked with the pendulum if there was any benefit to be gained from that quarter. Two round river rocks later, one on either side of the bed, negative magnetic polarity facing up at the right, for the positive charged hand, and positive magnetic polarity at the left, for the negative charged hand, a gentle source of energy was introduced. In probability, if a session of laying on of hands had been performed, the patient would have passed. My hands, like anyone who practises these forms of therapy, come in with a score of 70 (for the left) and 65 (for the right). The rocks were at 35 and 30. These measurements are taken from a chart of my own fabrication, there is no need for a measuring unit, unless one is wedded to quantitative science, comparison is a better standard, it tells you immediately if there is alteration or not. Incidentally, the patient recovered and went his happy way.

The next onion-skin of magnetism would logically be what we call rather obscurely polarity.

MAGNETIC POLARITY

The central position of the North pole in relation to its associated star, Polaris, can be readily appreciated by setting a camera at a very low speed with a long-time exposure pointing at Polaris. The pole star remains stationary and central in the image with all other stars forming circles around it. This would indicate a number of things, firstly, that the earth is not spinning at 105,000 km/h or so as we are told, otherwise the camera could not have remained focussed on Polaris, or Polaris is spinning at the same speed, which would be hard, if not impossible to explain. It further indicates that the stars and the planets rotate above our heads, thereby providing another proof of stellar and planetary movement in a circular fashion over the earth, so causing the dynamo effect mentioned earlier.

The magnetic attraction as exercised by the North pole has a definite phenomenal effect of drawing a lodestone, iron, needle or magnetised substance. From time immemorial, that property has been used, as it is today, in all navigation systems. This is not to say that there is a so-called magnetic force exercised by the pole. To be effective a dynamo effect must operate on a fixed axis, and that is what – I would maintain – is the line of force between the North pole and Polaris.

This phenomenon of attracting a compass needle is only implemented from the Northerly side of the direction in which the compass is held. The opposite end of the needle is not attracted to anything, it simply follows diagonally opposite the alignment, without bending or fluctuating movement as the needle would do if it were really subject to two attracting forces. The so-called South pole, however, can not reasonably even exist without bending the compass needle.

We are under a major illusion if we believe that the South pole of a compass needle functions under the influence of a power of attraction coming from that area, it doesn't. The sole power of attraction in a compass is coming from the North pole which draws the magnetized needle towards itself.

You must draw the logical conclusions. They are not, however, the subject here, despite the interesting ramifications of such a blatant fact.

To add to the confusion, when considering the reputed magnetic properties of the poles, we are told that in actual fact the North pole is pointing to the South, because those are the poles attracted to each other, and the North is indicated by the South-seeking pole. The complexities of this last statement are immense because that would imply a need to revise the entire paradigm with regard to gravity, fluids, gases and mass, let alone biology, geology and the behaviour of water. The magnetic influences will remain a mystery so long as we do not assume a wholistic approach of research and total environmental relationship.

Having said that, experiments have shown that exposure, for example, of earthworms to the South pole, rather the positive polarity of a magnet are deadly; after a three-day period, the worms were dead and dehydrated. Even worse was found when chickens were exposed to that polarity, they became cannibalistic and even attacked other animals, including cows! One can deduce that removing the harmonising influence of the north pole of a magnet can lead to trouble. Consequently, the balance is best ensured by limiting excessive magnetic influences. That is of course severely compromised with the generalised use of electricity, radio telephony and modern communications.

So, irrespective of the electron spin of the one polarity being in one direction only, things do not fall apart as a result of the repulsion.

On the contrary, due to their form of movement – in a spiralling vortex – they travel in both directions and stay together as a coherent whole. This can be seen in a relatively simple experiment with a slide under a microscope, a few drops of diluted sulfuric acid, or even better whole blood, in the slide with a magnet at either end of the slide. By switching the polarity of the magnets, the spin of the blood or acid will be seen to reverse. It can be compared to the Yin-Yang of Chinese metaphysics, but there is no need to create a non-existent zone and claim that it enforces a phenomenal energy. The danger of postulating fictitious fact whilst ignoring the operation and form of movement of energy is great, it is to be denounced.

Yet, apparently, we still do not understand how magnetism actually functions, and we are blithely informed that it can be explained away chemically as being an affair of paired or unpaired electrons. How sad to pass over with such nonchalance a living force which provides us with, or deprives us of life!

When we know that this magnetic force has been enclosed for billions of years in a relatively (if not completely) hermetic container, there is a strong chance that it will have created an environment which produces stuff. It has and it does, US and everything we know. Thanks to this mechanically simple but immensely complex mechanism of which we are a totally integrated component the whole phenomenon stays on track.

I would maintain that the ancients were very aware of this force and expressed the awareness in their fashion. We know, or rather can infer, that benefits were and are derived from stones and things made of stone. For the simple reason, stone can be a highly magnetic substance, with a negative and positive polarity, as seen above with the bed-ridden patient.

Consequently, and from now on in this essay, reference will be made to the positive and negative polarity – the Yang and Yin characteristics of material, stone or human. That will be to avoid the confusion associated with North and so-called South poles.

These are two totally different aspects of the phenomenon, and due to the lack of detailed research they are lumped together as if it were one subject. The power of magnetic attraction from the North pole is one thing. The verticity or spiralling motion caused by the flow of movement of energy in its multiple forms is another, and most probably what creates a positive and negative or opposing forces, although not necessarily the power of attraction and repulsion because they accompany each other at all times and in all places, and in varying degrees of strength.

That spiralling motion is found throughout Nature and is probably the most common form or pattern of the life-giving force on its journey. There would be every advantage in studying the individual forces in the equation of positive and negative, before categorically claiming the properties of one, while ignoring those of other energies at play, but they can be as subtle as they can be earth-shattering, so it is no easy task to establish quite what is at play.

It is not so hard and fast as science would have us believe, and definitely not as simplistic. We can impose our theories to our heart's content, but Nature continues her simple way. It is perhaps possible for us to learn, but I suspect only by assuming our natural birth-right, the role of observing.

PARA- AND DIA-MAGNETISM

Magnetism as it is manifest in our environment is considered for scientific purposes from its properties as we saw above. There is a useful and practical categorization with regard to the quality of force involved, namely paramagnetism and diamagnetism, but it is far from obvious as, once more, there are apparently no hard lines drawn in the sand. Philip Callahan has perhaps been the most instructive voice in recent years, drawing parallels in the natural world of insects and stone, connecting dots which have not been correlated before, in his useful volume, *Paramagnetism*, he fundamentally explains:

- Any substance (rock, soil, wood, whatever) which moves <u>towards</u> a magnet is paramagnetic. Yes, even wood has a mild magnetic capacity.
- Diamagnetism is simply a weak form of magnetism. Water and most organic substances are diamagnetic.

He is the only recorded modern scientist to have spent time investigating a round tower, especially the Devenish tower, near where he was based during the second world war. He even developed a very practical instrument capable of measuring the energetic force emanating from a tower. Unfortunately, that instrument has long since disappeared and awaits a resurrection. His work was ground-breaking and invaluable for the modern researcher, he did however miss a few fundamentally essential elements with regard to the towers, on which more later.

Once we move on from the limited 'one or the other' way of considering what is happening, and integrate Nature's multi-dimensional system (which we cannot avoid) with her inimitable fashion of blending the two extremes, no matter what we call them – *yin/yang*, male/female, north/south, negative/positive, life and death, we can approach harmony, we are able to emulate Nature when we work in her service, along the lines of her patterns, and that is what, I would maintain, the ancients were doing with their stones, the pyramids, the stone circles and alignments, and of course, the round towers in Ireland.

MAGNETISM AND WATER

The relationship between magnetism and water is quite obscure, if not non-existent from the modern scientific point of view, whereas when considered individually the connection might appear more obvious. As indeed it is when considered wholistically. Given that our planet is made up of about 73% water, like the human body (what synchronicity!), it would be logical to work on the hypothesis that water is a mediating factor. Especially when we know that water is (somehow) claimed to be the dedicated medium for the 'magnetic force' from the moon, and the essential component in the alchemy of vital force from all sources of life as we know it. The movement of water in the atmosphere, on the surface and below ground make it the prime mover of all the elements on which we depend for life here on this planet.

It would require several volumes to do justice to water, her life development, her capacities and functioning, let alone the mythological references. One thing is quite apparent, however, and that is the awareness of the ancients concerning the importance of water and her abilities, as we can surmise from the vast pantheon of gods and goddesses of water, not only in the skies, on the earth, but also underground. There is also a plethora of mythical beings related to those aspects of the combined effect of water and her surroundings, dimensions that were as strange to the ancient observers as they are for us now. A few examples of those beings would include the wyvern, naga, dragon, manatee, and so on.

If it is unknown today precisely – or even vaguely for that matter – how the human reacts to low-field magnetic forces, one can project what might have been in the minds of ancient humans, and the imagery they devised to deal with the inexplicable, but oh so real, forces that have not changed in manifestation since time immemorial, even though somewhat weaker now due to our incessant tampering with Nature.

What is more, the pre-Christian traditions in Europe, and animist cultures elsewhere in the world all give importance to certain zones where water rises to the surface such as springs and wells. Frequently, these sites were the subject of considerable local and more distant attention, with pilgrimages being made to the sites that often became shrines, widely respected and protected by their reputation alone. These springs were, and still frequently are, venerated for their special healing aspects, qualities or inspirational properties.

Naturally, whenever a new faith arrives on the scene every effort is made to eradicate the previous beliefs, especially in the western world, early 'Christian' practice was extremely militant in doing this to the preceding traditions, especially the pagan Druids, Irish, Welsh and Celts. It was soon apparent, however, that destroying the sites would cause adverse and hostile reaction, so what better way than to incorporate the former belief into the new religion, and all parties are content as time works slowly but surely at installing the new faith. The problem of course is that the *raison d'être* of the site is slowly forgotten, as are the technical details of the how and wherefor of the technique employed. What we forget is that has seemingly made no difference to the ability of the water and its companions, and the miracles still occur.

Consequent to the removal of the natural and practical reasoning for the original setup, all explanations as to the importance of underground currents are eclipsed, and soon replaced by some article of faith, which completely blurs the understanding of something so simple as a geological fact. This then becomes the source of the 'magic' which will still change with all the factors that affect hygrometry as we know it now. And that is something we tend to eclipse, despite it being quite easy to prove with a few simple experiments.

In a manner of speaking, water stores energy, a fact recognised today by authorities like Gerald Pollack, as it was in ancient times, in one of the oldest Vedic texts, the *Chandogya Upanisad*, it is said "Water is *prana* (life-force or vital energy)". We had to wait for Viktor Schauberger, however, to inform us that this is probably very dependent on the quality (or maturity) of the water.

What a long way we are from the modern, sterile concept of H_2O!

There is a very tight relationship between the forces on the surface of the earth and what is in the earth. We have a very limited nomenclature and definitions for these energy forms, and Viktor Schauberger is surely the only person in recent times who has not only drawn attention but has even tried to explain and connect the actual function of these natural forces. The fact remains that when one modifies the magnetic influence on the surface, by laying out strongly magnetic round river rocks for example, the neighbouring unfolding materialization of magnetic fields is altered.

VIKTOR SCHAUBERGER (1885 – 1958)

As you know, there are certain people, incidents and things in our lives that leave indelible influences. Schauberger was one of those for me. His simplicity, perspicacity and candidness are shining examples, what is more he shared his knowledge and the fruit of his observations freely.

One particular incident from his forest warden days in upper Austria that he recounts is especially important to this narrative.

In a region of the high mountains where water was scarce, was a spring covered by a dilapidated stone dome that Schauberger thought was in danger of collapsing. His older hunting buddies warned him of the danger of the spring drying up if the stones were removed. Attentive but sceptical as always, he had his team number the stones as they dismantled the structure. Sure enough, the spring ran dry, so he rebuilt the primitive structure in the precise order of its disassembly, and three days later the water was turned back on. What happened?

I mention this simple incident because it illustrates a major unknown relevant to this essay, namely, the unsuspected properties of stone and its interaction with water. Schauberger's form of observation was wholistic, focussing on water because he not only knew how important that element is to life, but he played with it. He provided insights into the intelligence and capacities of water as no one else has recorded to date.

There is a strong argument for the use and application of stone when there is a scarcity of water, or, as in the case above, when there is a need to encourage water to the surface for reasons of human settlement, agriculture or maintenance of the harmony that water in the right quantity ensures for all of Nature.

The universe of magnetism has much to teach us, but we can not expect any help or explanation from the successors of those who usurped those skills from the original observers of Nature. In itself that may not be a problem, but great damage is done when the understanding or the thorough gnosis of the situation is eroded by the passage of time as a result of the deliberate obscuring of the origins.

In my professional capacity as a technical translator in France for some twenty years, I worked for several large water companies, and even had the chance to visit some of their installations. Hands-on experience of the technology employed was hugely instructive, but sometimes equally horrifying.

In all probability, the public authorities have no other solution but to add chlorine and fluoride to the water supply, but not only is that extremely damaging to human health and the environment, but it complicates the purification process of water when it is recycled.

For example: Water is recycled when collected from the waste disposal facilities that equip most large conglomerates, it is then processed, treated and redistributed into the mains water supply system. In other words, water is consumed, urinated, recycled and consumed once more. The purification methods generally used eliminate many of the nastier bugs, but do little or nothing to the traces of oestrogens, recreational drugs or antibiotics that remain in the water AFTER the various processes.

Release of the "purified" water occurs on successful completion in the final tank. A fistful of very small crustaceans is thrown into the water, if these creatures scarper across the surface of the water, out the water goes into renewed service; if they expire, which they will do rapidly if the required conditions are not met, the water goes back into the purification process. A rather primitive but totally prevalent and standard procedure throughout the world.

This makes it easier to understand the purely chemical approach to water, as opposed to the wholistic, life-giving function. It would also provide the basis for meaningful research into many of society's modern ills, if we are consuming a potentially regular dose of chemicals, already assimilated by other humans and animals. That research is for another day, as are the effects on humans, animals and agriculture, but could go a long way to explain the increasingly common complaint of intestinal fungal issues – the gut bacteria are no longer subject to the conditions essential for their existence, and the period of adaptation risks being long, if even possible.

Water is multi-dimensional in its scope of action and influence, with an intelligence of its own and memory being a component of that intelligence. Jacques Benveniste lost his name in 1988 for stating something similar, but it seems that we are at last opening our eyes to this reality, with mainstream science even accepting Pollack's fascinating work – *The Fourth Phase of Water.*

However, let us return to diamagnetism and paramagnetism. From the broad, as opposed to the atomic perspective adopted by science, the subject is perhaps worth considering by looking at what happens to water in the open seas. It evaporates, we are told, but not of the exact details how. I would venture that surface water is less diamagnetic than the water at a lower level, most probably due to the temperature difference – the seminal 4° C as demonstrated by Schauberger – and is able to levitate (despite gravity!), attracted by the more paramagnetic environment found in the atmosphere. Once more, an example of the very weak field magnetism playing an essential role in governing life. Nature is the most amazing manager, so I can only hope that the effluents from the land, laden as they are with chemicals and irradiated substances, prove amenable to her capacities out in the open seas.

It is a very subtle mechanism, as any gardener will tell you, a soil boosted with paramagnetic supplements (e.g. volcanic soil) will generate a richer crop than one deprived of the same. As Callahan elegantly points out in his book *Paramagnetism*, old-school Japanese gardeners are very aware of these properties and dispose the plants in their gardens according to those principles. The reason being that some plants profit from the presence of paramagnetic stones, whereas other fare better in the vicinity of diamagnetic stones.

There is every reason to take his rationale a step further, we are dealing with the positive and negative magnetic polarity of a stone. That is a fact which few people are aware of. A stone/rock/boulder/menhir/dolmen possesses magnetic polarity, hence the notions of male and female, yin and yang associated with stone. A simple nomenclature for sure, but of significant relevance.

Because certain types of stone achieve substantially different forms of energetic output.

That statement involves two aspects which I would like to clarify and which, to my knowledge, have never been mentioned before, either in relation to stone, or to any of the other stone applications around the world. Both of which are significant, because key to all erected stonework considered in this study.

Given that a substantial area of the earth is made up of water, it is likely that the component anions and cations in salt-water have a knock-on effect, due to conductivity or straightforward sharing ability, with the neighbour, rock. When we take a closer look at a rock, in order to establish whether it has what I refer to as magnetic polarity, it is invariably the case.

This subtle, and neglected – because generally unknown, property of stone is its magnetic polarity. The ability to arrange the components of a stone into a magnetic structure which can then be recognized by us as such, is in all probability the influence of water, or fluidity, its ionic exchange and varying temperature. The idea here is not to investigate scientific cause and effect, even if we were able to deduce the how and the why, rather, what use can be made of this understanding to enhance our lives and benefit the natural environment, as there is strong evidence to believe that this is what the ancients did. They used stone to adjust the movement of energy so that one and all benefitted. So, let's keep to that pragmatic approach.

THE STORY OF HOW THIS ALL CAME ABOUT

Some fifteen years ago, I was completely unaware of this strange capacity of stone. I was teaching (Principles of Translation, Religions, Academic Writing, Buddhism and Science) at Mahachulalongkorn University in Chiang Mai, established in Wat Suan Dok, a school exclusively for Theravadin Buddhist monk students of humanities from Asia, mainly India, Nepal, Bangladesh, Burma, Thailand, Laos, Cambodia, Vietnam and Korea. One day, at the end of an Academic Writing class, which no one could give a damn about, I was explaining the notions of Shankara's kevala-advaita philosophy to a group of mainly Cambodian students in their fourth and final year, when a pale-face monk walked in and asked to join the class. At the end of the session, he asked if he could join us the following week as he had mistaken my class for the one that he wanted to attend, but found the subject much to his taste. He was a recently ordained Polish monk living in Wat U Mong at the foot of Doi Suthep, Chiang Mai's sacred mountain. For many years, Phra Chris (PC for convenience) had lived on the island of Skye where he had been befriended by a German radiesthesist, and had learned a great deal from this clearly erudite man, who I would mention in passing had started his medical career as an army doctor, survived and escaped from Stalingrad, and ultimately retired to Scotland. This new-found acquaintance from among my Buddhist monk students agreed on an exchange of services. My knowledge and experience of laying on of hands and fasting, for his knowledge of radiesthesia. So started an intense, three-months of experimentation and research.

We both shared a keen interest in sacred and applied geometry, alternative medicine and techniques, but especially the practical applications that such arts permitted. He was an accomplished dowser, and he taught me – a ready and natural subject, what he knew. Our main focus was on the use of geometric patterns and forms in our large garden in Mae Ann, just outside of Mae Rim, in the Himalayan foothills.

One day he mentioned in passing that there was a large pile of rounded river rocks in Wat U Mong, and that the abbot of the monastery had given him permission to use them. He was of the opinion that river rocks had some special property and was keen to experiment with them. We collected thirty or so large stones, small boulders weighing 15 lbs or so, and brought them back home in my pick-up truck.

Using a pendulum and a Bovis scale chart, ranging from 0 to 40,000, we started recording our findings. Initially, the score of the individual rocks to be used would be written down; then, once placed in the desired zone, the reading on the scale would be recorded. By carefully measuring out distances with a measuring tape, the rocks would be placed in various geometric forms (squares, circles, rectangles), including layouts incorporating values such as pi (3.14159) and phi (1.618). We were intrigued by the results and substantial distances that the influences of these patterns carried. There was, however, a strange phenomenon that was causing some confusion in our minds.

Even though using exactly the same measured distance when laying out an identical form, a square for example, we were finding readings with a difference of several hundred units. Was it the ground (the local earth has a heavy component of metal), the presence of trees, human presence,...? It was a tough nut, and of course, it was more by accident than logic that I discovered one day that the same rock was giving a different reading as a function of which side was being measured.

We did not have any magnetometers, and even a meter which I later acquired, did not give any indication of what we knew was a subtle, occult property of a stone, so pendulum and a chart it was.

By simply turning the stone over, the reading was found to be substantially different, a matter of a thousand points or so.

As a pendulum can only provide a binary answer, a yes or a no, a positive or negative. That is when it occurred to me to ask, 'Is this positive?', not knowing quite what the connotations were to such a question. One side of the rock produced an affirmative reply. So, second question, 'Is this negative?', the same result from the opposite side of the rock. What is the score on the 'negative' side? 15,500. On the 'positive' side? 13,500.

An initial understanding dawned by applying an apparently logical approach: if the same rock gives two readings, one side higher than the other, the effects of the higher (positive) side can potentially produce higher aggregate scores from the whole layout when measuring the reading on the chart. Would that provide empirical evidence, at least, that we were on the right tracks?

By repeating the same patterns, but with the rocks all negative-side upwards, the readings were off the Bovis chart. For those of you interested, you will find in Appendix 1, some recordings of our experiments.

A word here about the Bovis chart might be in order. In about 1930, a boilermaker by trade, André Bovis from Nice in France, devised a system to measure the vital energy of food, specifically the fruit and vegetables from his garden. He observed that their nutritional energetic value diminished over time from the moment they were picked or cut, especially the leafy vegetables. He created a scale, which he named the Biometer, with a measurement spectrum ranging from 0 to 10,000, calibrated in units that he believed corresponded to angstroms, and then recorded his findings using the score indicated on the chart by a pendulum held in his hand. This scoring unit later became referred to as Bovis units in circles interested in physical radiesthesia (the French school of dowsing whose proponents tried hard to get dowsing accepted by the scientific community as a valid means of research and analysis in the field of physics).

Consequently, PC and I designed and drew a new chart, this time from 0-100,000, we took new readings and found some very elevated scores, 45,000 to 60,000. Spending time in such 'high-energy' environments clearly had an exhilarating effect on us, but it was only when my wife remarked that my hair was standing on end (PC, as a monk, had a shaven head) and both of us were especially bright-eyed, that we became aware of the altered state we were both in, after working in the neighbourhood of these rock formations we were creating in the garden.

To resume our findings and propositions, we agreed that one of the special characteristics of a river rock is its 'magnetism'. Perhaps it has assumed the round form as a result of being rolled in a river, or of the water flowing over it for many years – perhaps even hundreds of thousands, probably resulting in their picking up a charge, for the readings from one rock to another differed, and rightly or wrongly, we assumed and called this the 'magnetic charge' that we were picking up on with the pendulum and Bovis chart.

There were a number of intriguing phenomena discovered as a result of setting up these 'power patterns'. Firstly, was the shifting of a point where lightning was in the habit of striking. This spot, beside our guest house was clearly defined by the lack of vegetation because of the regular scorching that the area of 4 square metres or so was subject to. Northern Thailand is not especially bothered by lightning although there are numerous storms in a year, however, after placing the stone circle, some months later the strike zone was no longer there as the area grew grass anew, and that still holds true to my knowledge.

The second event was the appearance of what I can only describe as arranged energy points, valid so long as some of the more powerful patterns were in position. In the process of checking readings at a distance from the centre-point, we measured distances in paces (metres) and at given angles. While doing that, it appeared that there were areas with higher readings regularly dispersed every fifteen metres, seemingly in a well-defined pattern, running east-west. The individual point was showing a consistent reading of 20,000 over an area of 1.5 metres, and that over an extended area. That extended zone had its limits when a new telluric force (underground stream or cavern, geological fault) manifested on the surface, or topographic incidents occurred, lake, slope, etc.

What happened next for me was a period of intense study into all subjects related to magnetism, physics, radiesthesia, Egyptology, and what is rudely termed alternative medicine in a concerted effort to discover if there was further information to be found regarding this natural phenomenon and its possible use in the past. Over the next few years I read a great many books on these subjects, from the classics in the genre, to the mavericks. Nowhere could I find anyone who shared this appreciation of stones having a magnetic charge, nor what could possibly be done with them, despite the ill-defined allusions to subtle energies.

PC and I went our separate ways. He defrocked a year or so later under a cloud, but he was a great inspiration in the time I knew him.

My interest was seriously stimulated, for it is not only rocks that share this 'magnetic charge', but people too as I was about to discover from the works of Leon Eeman, Mesmer and von Reichenbach. It was not as straightforward as might appear, because not only do the forces in our environment change, but they have an intimate and immediate effect on our metabolisms. We neither recognize that, nor actually see it happen, although we may suffer physically but do not correlate the coincidence.

We all, human, animal and plant have feelings. Perhaps even inanimate objects too, I wouldn't discount it. The majority of people are also aware of their surroundings. They 'perceive' things, with all 365 or so senses, not just with the five Aristotelian senses (more to follow on this subject). Now while they may not be able to say exactly what it is they are feeling, they know if it is comfortable or not.

If it is an influence which causes discomfort, they are sensible enough to remove themselves, when possible, from that troublesome source. What is more, sometimes, they might even be able to determine the cause of the unease and change certain factors so that the problem is resolved, and harmony is restored.

Well-being, harmony, comfort are all so many expressions that convey what all beings on this earth are after. That much seems clear, and it would be reasonable to assume that has always been so in the known history of human existence.

Without going down any rabbit-holes of metaphysical speculation as to what this balance is and where it lies, can we simply assume, along with the Chinese, that when heaven and earth are in harmony, there is the environment suitable for human existence?

A basic enough place to start our investigation, but which fits the bill to perfection.

The importance here lies in the fact that this sensation of health or well-being is facilitated, if not determined, by the synergy of what is happening in our environment, which of course, can only be brought on to the radar-screen of our awareness thanks to our senses. A highly complex process no doubt, but which depends entirely on the right balance and/or maintenance of temperature, air, water, light and movement, etc. A very delicate act which can come to a grinding halt if any of the components are modified to excess or depletion. This has apparently always been the case, and even though there have been a few wipeouts in the past, we muddle along in those narrow parameters which separate life from extinction, in oblivion as to what is actually happening, let alone how it actually functions.

During those periods of muddling-along, humans have devised and created some remarkable ways to improve our lot. The common denominator in the methods seems to be the interface between the individual body and the external world, and *vice versa*. It is rare that man tries to benefit the environment by working the inverse, i.e. external to the individual. This is an idea that will be explored here, because the unsung hero workers of stone seemed to be doing that.

But what exactly do we know about how energy-exchange really works? Rather than start a polemic which is unlikely to be resolved, can we assume a wholistic approach, an overall, non-differentiated, interconnected and communicating view, as much as that is feasible for a simple human to achieve?

There is a dearth of information concerning this aspect of energy in our western cultures, but that is not the case for other current and older cultures. There seems little doubt that such a state of affairs is deliberate on the part of those who govern us, otherwise there would be no reason for the restrictions as there are, for example with regard to frequency medicine, practised almost uniquely in China, and locally at that, because elsewhere it is illegal. But that does not help us progress in our quest.

If energy, as expressed in the living organism or the body, is understood as the vital force organizing the flow of blood and fluids, breathing, the nervous system and the vital organs, all operating thanks to a constant flow, we not only find affinity with the theory of the meridians in Traditional Chinese Medicine (TCM), or the three *dosha* (humours of wind, fire and mucus) in Ayurvedic medicine, but there is strong evidence that the ancient stone-masons had a firm understanding of the flow of energy in the earth. The common denominator is movement, no matter the speed.

Energy can only act and interact when movement occurs. Death is the absence of movement.

Modern science has done a grand job of differentiation, determining the potential from the kinetic, introducing the six basic families of energy, thermal, chemical, electrical, and so on. What is very encouraging is that there are even some schools of modern physics now coming round to the idea of an interconnectedness of these different forms, either as a coherent totality or from the position of individual consciousness. This is wonderful but somewhat like reinventing the wheel for those familiar with thought systems from older civilizations; however, better late than never.

We moderns have still not yet grasped how energy 'operates', I would venture for the simple reason that the emphasis is constantly on the physically apparent manifestation. Little or no attention is given to the subtle processes that go on invisibly in those parts where we do not currently look.

For example, in stone.

Energy is organic — it is living, whether that is apparent to the senses or otherwise, kinetic or potential. Everyone seems to agree with that, and you might add, "And so what?" Every thing is formed of energy, and is therefore alive. There can be no exceptions to that law. We just cannot understand with our reasoning capability what energy or life is. What could be more natural? We are a part of that life, our physical bodies are formed of materialized energy so to speak, and our spirit or vital force is energy personified, the aggregate of thought necessitating a physical entity to accomplish its desires. It would be inconceivable for the energetic entity of the human to somehow position the sense organs into observing and analyzing their own source by means of their own manifestations. But to have an intuition as to what is going on is another matter, because we are no longer restrained by the trammels of the static cerebral intelligence which tend to be rather static.

If we can put aside all considerations as to quite what we are as individual human beings, and consider a broader picture where Consciousness apparently animates a complex energetic mass that we neither control nor comprehend, then we stand a chance of catching a glimpse of reality. This compound of Consciousness, thought, intent, body, and will, (especially thought), seemingly differentiates us from the rest of "animal" existence, and that difference is probably what makes us think we can understand it, simply because we would like to.

We share the same energetic properties as plants and animals, but is it not this force which enables existence? Is that not reality in its most pristine form? Never to be perceived, but always present. And existence shares the characteristic of consciousness and well-being in the Upanisadic teachings.

It would seem apparent, however, that this sense of well-being can only come from serenity, and as Laozi stated back in the 6th century BC, that is a state of mind available to one and all. Obviously only a few people practise that way of being, and as a result, it appears exceptional. We in the western world have learnt to differentiate psychology and medicine, whereas for the ancient Chinese these two were the same thing and serenity can be succinctly summarised as the control of emotions, or rather not so much control, but more not being disturbed by them. This is achieved by being constant, as it is for the Vedanta of the Hindus, and its accomplishment is illumination or realization of what's going on.

From a modern materialistic viewpoint, constancy is considered something both boring and dull, even to be avoided. That wasn't always the case. In the environment we know here on earth, there is nothing quite like stone to represent constancy, and it would appear that our forebears appreciated that the equilibrium needed to be in tune with our destiny was embodied in stone, what is more that equilibrium can be enhanced by the use of the inherent magnetism of stone. The use of stone for human habitation is a relatively modern phenomenon, evolving at a rapid pace in the last few centuries from wood and earth, to the predominant medium of concrete. The monumental structures found throughout the world no doubt served a variety of purposes, which are beyond the scope of this research, but generally speaking, were for the benefit of one and all in the immediate – and not so immediate – community. These structures were not used for individual accommodation.

That elusive balance, sought after by some, discarded by others, has in all probability not evolved over the ages, whereas it is more the life-style which decides the method, and the means. That is what makes it difficult for someone of another time and space frame to comprehend the fundamental motivations of those from an earlier epoch. To all intents and purposes, modern life (especially city-dwelling) employs an utterly different mindset from the one used by people required to live according to the dictates of Nature, not only as regards life-style but with regard to priorities. In the distant past the emphasis was perhaps more oriented towards unity rather than self-fulfilment, an association with Nature rather than random alienation, and vicarious satisfaction.

Over the course of time, Nature has allowed/inspired humans to devise and create some remarkable ways to improve our lot. The problem for us today, is to imagine what the purpose might have been in earlier times, and presumably our ancestors were not taken with a sudden urge to build stuff. This is the idea that will be explored here, because the unsung hero workers of stone seemed to be doing something with a specific idea in their heads, when they deployed them for the benefit of society.

As is often the case in human tradition, we either ignore or forget the original purpose of objects used in our daily lives. This seems to be the case with standing stones, whether menhirs from megalithic cultures in Europe and America, the *lingam* in presumed phallus worship of Hindu origin, the supposed territorial marking stones used in SE Asian architectural sites (*singa* or *sima* in Pali, *sema* in Thai, *srnga* in Sanskrit), the pyramids of Egypt, the round towers of Celtic Ireland and Britain, the moai of Easter Island, Hindu, Buddhist or Khmer vestiges. It is of etymological and intuitional interest to note that the Sanskrit word, srnga, means horned as in a cow. The cows' horns act as antenna. Excuse the red herring!

In our modern age, it is the results-based criterion that carries the day, so applying that same model to the subject here, I propose a closer look at the physical structures in question using a different means of analysis and understanding, which despite the current context where there remains an air of academic secrecy, logic would indicate that there must have been a clear purpose at the time of their construction.

Before we get there, it would be in the natural sequence of things to look at magnetism in the body first.

CHAPTER 4
MAGNETISM IN THE BODY OR BIOMAGNETISM

Water in the form of sap flows in a tree, up during the daylight hours, and back down during night-time. Water in the form of blood flows through the veins and arteries of mammals, also following diurnal and nocturnal patterns. Other vessels carry nerve impulses, and the meridians convey their energetic flows. These amazing (because complex life-providing) processes occur for the duration of physical life, requiring no effort, nor cognisant control.

What is remarkable is that we pay no attention to the magnetic charge these flows generate, because there can be no doubt that they do just that. The resultant magnetic aggregate is surely a major constituent of life, and merits the epithet of biomagnetism. It is a significant health factor in frequency medicine, and of special importance nowadays when out biotope is totally polluted by electromagnetic frequencies of the most unnatural sort. I would even venture, on the strength of my work experience, that maintaining a strong magnetic field is key to good health.

It is intriguing to find that this quality of magnetism, is present even after the fall of the physical envelope, inasmuch as the magnetic charge or field of force continues its presence at a constant, albeit at a significantly lower rate than when living, namely at a score of 15. This could perhaps indicate some form of remanence relating to existence, never lost but simply transformed. Whatever is going on, is politely ignored in medical science – we measure the electric current of a patient to reveal reputedly important factors like heart function, but elect to sideline the associated information, and remain in complete unawareness as to this aspect of life and its components. The development of this idea is part and parcel of the theory being developed here, so please bear with me.

This 'something immaterial', the magnetic, or biomagnetic field of an individual, can be an immensely useful indicator of human health, and ranges in measurement from 15 to 75 (this is on a chart of my fabrication); there is no unit of measurement *per se* and no registering of the electron spin or other movement.

The unit of measurement used to record the magnetic field of the earth is a microtesla, ranging between 25 and 65 microteslas for the earth's charge, I suspect that this same value can be applied to a human and is what I am picking up with a pendulum on my chart. Quite what this expression of movement is can be explained in a variety of ways, but I don't think that really helps our understanding. It is probably a function of the life energy, the Chinese *qi* or the Vedic *prana*. The strength of such an element in a living human varies according to many factors: exposure to electromagnetic radiation, the strength or otherwise of the immune system, the wearing of rubber-soled shoes, environmental influences, physical blockages, emotional well-being, and so on.

It can be strengthened, and that is very good news because our entire environment is gradually being weakened magnetically.

It is useful to know the magnetic polarity of a person's hands if you are working on them in person, for it seems best to work on the positively charged hand. Generally speaking, men have a positive charge emanating from the palm of their right hand and a negative one from the left. Men who were born prematurely or who are gay often have an inverse polarity, making the left palm the positive and the right, negative.

Women, in contrast, have a positive charge emanating from the palm of their *left* hand; women born prematurely or who are lesbian may have inverse polarity, and almost invariably at menopause their polarity switches, the right hand becoming positive and the left negative. This last observation might explain the severe changes to women's hormonal system, as experienced with flashes and other uncomfortable effects. It is not an instantaneous reversal but a gradual process, often taking several months, even years, hence the inconvenience and discomfort of menopause lasting in some cases for several years. The condition of being right- or left-handed, even ambidextrous, seems to make little or no difference to the biomagnetic polarity of the palm of the hand.

The images below shows the locations of general biomagnetic points in the human body as explained above. As you can see, there are a number of constant points shared by both sexes – the forehead is positive, the back of the head, negative; the top of the spine, positive, the base, negative. One negative point, the navel. That is perhaps no surprise, as it is the source of life from our mother. In the *Caraka Samhita*, one of the principal reference books of Ayurveda, the navel is known as the *nabhi cakra*, incidentally the only time the Sanskrit term *cakra* is mentioned in a medical text, as opposed to the much later Tantric texts (about 16th century onwards).

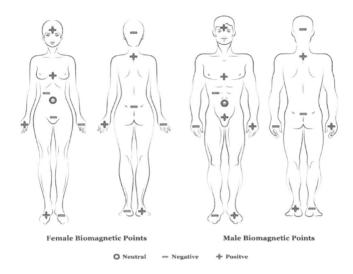

Female Biomagnetic Points   **Male Biomagnetic Points**

O Neutral   — Negative   + Positve

With the same constancy, the back of the hand will have the opposite polarity to the palm, irrespective of the polarity of the palm. If the left hand is positive, the right foot will be positive too, and vice versa for the right hand and left foot.

The naturally harmonious, positive and negative arrangement of biomagnetic polarity in the human body is generally constant for the duration of a lifetime. There are circumstances in which certain aspects of the arrangement can change, most notably for women at menopause, when their biomagnetic polarity changes. This fact is patently ignored by modern medicine, perhaps most other systems too; however, there is some relief to be found in magnetic recharging and making an effort to raise the magnetic charge.

This involves applying Leon Eeman's simple exercise that allows you to recharge your biomagnetic field in just a few minutes. Regardless of the polarity in the hands, interlace your fingers, and let your hands rest on your stomach or chest. The polarities will be joined, negative and positive. Now, your feet. Those with right-handed positive polarity, put your left foot over the right ankle. Conversely, for left-handed positive polarity, put the right foot over the left ankle. You will find out after five minutes if you made the wrong choice. You will feel enervated, so change the position of your feet.

With hands and feet joined, sitting at your desk, on the bus, lying down, wherever, you have closed the biomagnetic loop of the body's field and are recharging. Why do that? As explained above, the current of the blood's movement through the veins and the energy through the meridians result in an electric charge, which can be measured in millivolts. Ten millivolts, and you are at death's door; at seventy millivolts—and you will not get higher—you are fighting fit. Most people come in at twenty-five millivolts, which is weak, further indicating that the immune system is probably struggling.

No one, to my knowledge, has ever pointed out that this is just what you are doing when sitting cross-legged in meditation, one hand in the other, but no matter. One often sees old people in this position; maybe their bodies react instinctively.

Contrary to what many people maintain, the polarity of human limbs can change, but that may be a recent phenomenon due to the surge in the past one hundred or so years of electricity, mobile telephony, wireless devices, personal computers and other electronic gear, air travel, the earth's weakening magnetic field (due to the extraction of oil and minerals), and so forth, likely resulting in the perturbed magnetic patterns commonly encountered nowadays.

Thankfully the immensely useful therapy of magnets, or their application, has been generally neglected, and as such there is no branch of medicine that is concerned uniquely by magnetism, so it is perhaps quite deliberately ignored. That is reason for celebration because it would most probably have been monetized and perverted. So once more, this technique offers some very useful ways to take care of ourselves, and consequently ensure a healthy level of biomagnetism. The problem for humans is that the original level was maybe not as high as it was for stones and plants, and we subject ourselves to greater exposure from electronic gadgets and electricity. There is no surprise to see a plethora of modern diseases arriving on the scene these last hundred years or so.

This digression into human magnetism is to introduce the idea of interconnectedness which we might suspect, but do not act upon. I would postulate that one of the purposes of the ancient use of stone and its placement was out of concern for human health, for our forebears discovered that there was a connection and, even better, a readily available application. One is also able to recharge magnetically from a stone or stones. This is where the more powerful standing stones especially come into their own. There are probably a variety of ways to accomplish this, by lying on a flat stone, by standing against or between standing stones, and so on.

It is impossible to determine all the effects at work in our environment. Let alone which ones are impacting us, because not only are our metabolisms different, but our magnetic fields are too. There are so many potential factors at play that it is positively daunting trying to calculate what is going on.

We know that the basis of medical diagnosis is symptom-based rather than frequency-oriented, and there is nothing wrong with that as such. What is wrong is that the remedy is systematically provided by purveyors of chemical compounds, irrespective of the initial observation and the observer's experience.

That is the way things are and there is little one can do about it, except go maverick and refuse the pill. Happily, more and more people seem to be doing that by searching for alternatives; and there are many valid options available.

Fortunately, however, some people do pick up on what is going on around them, including underground, but not many know of the potential issues, and even fewer, know that you can actually do something about it once you have discovered what that influence is.

It is a matter of sensing – one of our 365 senses I would add – because you are subject to the specific influence more than the next person as your particular metabolism works like that. However, unless you are familiar with Nature's ways, there is very little chance that you will associate what you are feeling with what is going on below ground, or the manner in which that energy form moves.

In the same way that heat and hydrogen rise, those energies ascend from below the ground and head upwards to the sky, no doubt in the permanent process of what we know as life. The recycling, regenerating effort that is in constant movement all around and within us. For that reason, and on the basis of the observations of past researchers from numerous cultures and traditions – but especially the Chinese and Indian, because we have explanations in their literature on which we can fall back to gain a better understanding, we could do worse than to pay closer attention to the archaeological vestiges left behind by our ancestors. There are in all probability messages which have been deliberately occulted in man's never-ending effort to impose control over his fellow-man.

Over the last couple of hundred years several very smart people have expressed some remarkable ideas on the subject of magnetism, as is only to be expected from normally curious individuals attempting to perceive the workings of the subtle life force. The work especially of:

- - Franz Anton Mesmer (1734–1815) who observed that all points on the left side of the body have the opposite polarity to those same points on the right side. He believed that the poles can be changed, communicated, destroyed, and intensified.

- - Baron Karl von Reichenbach (1788–1869) who found that the left hand has negative polarity, as does the *od* (vital force) in that hand, in both men and women.
- - Hector Durville (1849–1923) who maintained that negative polarity is to be found on the left side of the body, likewise for both men and women.
- - Leon E. Eeman (1889–1958) who found that both men and women had negative polarity on the left side, unless they were left-handed, in which case the polarity was reversed.

So, serious experimenters ranging from the eighteenth to the twentieth century have different things to say regarding human magnetic polarity. From this evidence, something bigger is clearly taking place, but no one in the scientific community – who might have the wherewithals to conduct a thorough research seems overly keen to investigate further. While it is a measurable fact that the earth's magnetic field is weakening. It would therefore be logical that this has a subsequent impact on humans.

If we are to find answers to these life-changing issues, it is essential to structure our research because it is apparent that the different researchers above did not ask the same questions during their respective experiments. In all probability, no questions regarding polarity were asked, it is only too natural to take an accepted fact for granted, even if that fact is incomplete.

In my practice when working with humans, I systematically question:

(a) what is the magnetic polarity in both hands,
(b) what is the individual's magnetic charge (in microtesla or other scale),
(c) what is the individual's electric charge (in millivolts),
(d) what is the individual's degree of vitality and general health (a suggested scale of 0 to 10),

(e) what is the individual's emotional/mental state.

Obviously, this is not an exhaustive list, and is very specific to my knowledge and understanding, but you get the idea of what could help ensure more coherent and consistent data. In all probability, the early researchers asked their subjects no questions at all because they were simply not familiar with the possibility that these factors can vary, and they were perhaps even less familiar with modern scientific methodology, even if they believed there might have been some merit in interviewing their subjects to gather as much data as possible.

Confusion over the centuries concerning magnetic polarity in the individual's hand could have been avoided if each person had been checked before, during, and after the experiment. Because our understanding is deficient in these matters, as is only natural if we do not recognize the importance of biomagnetism in the individual, it could be beneficial to revise our way of looking at things, which, as we know, develops, evolves, and sometimes regresses.

We can clearly draw no conclusions from these past experiments because of the lack of recorded parameters and it does not really matter, but in my experience it is definitely worthwhile asking on each occasion, with a pendulum or some type of muscle testing method, the questions above. The causes, as always, are beyond our ability to explain with any confidence, but one thing is for sure: if you don't ask, you are unlikely to realize that you don't know.

As I said in an earlier book, we should be aware of falling into the habit of dogmatically maintaining that a given problem can be worked out in precisely a certain way, as is so often the case today, with endless clichés about one practice being the "best" and another the "worst," don't eat this, eat that, avoid this, embrace that. It would be naïve and even impossible, even for an expert in the relevant field, to hold that a given practice is unequivocally better than another. Each individual will respond differently to the same set of conditions. If we all had the same metabolism, allopathic medicine would work across the board and good health would be guaranteed for all, but unfortunately that's not the case.

Yet for some reason we tend to accept solutions without question if we think we're getting a fair deal—we convince ourselves that we can benefit, if only we accept certain conditions. But it is those conditions that are debatable, that might not hold up under rigorous intellectual reasoning. In our haste to find a solution, we've stripped intuition out of the equation, and it's only in our heart's intellect that we can hope to find a true solution; it is our heart that has a connection with what is, rather than what our head dictates. We know where pure rationality takes us, and how easily it can be manipulated to the advantage of a few.

INSIGHT INTO MOVEMENT AND ITS PATH: THE SPIRAL IN THE HUMAN

Some fascinating research was conducted in France during the 1930s by two women, Jacqueline Chantereine and Dr.Camille Savoire. They suggested that the human receives telluric and cosmic energy via the body along well-established pathways. Their conclusions led me to experiment further along the lines of biomagnetism.

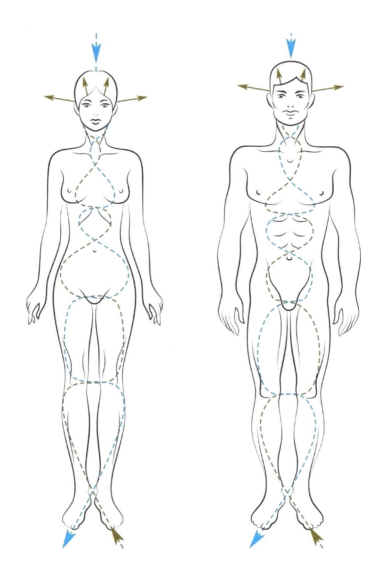

Cosmic and Telluric Force Entry and Exit Points

Chantereine found that the telluric *yin* energy enters the body through the left foot. She found that the energy rises in a spiral through the body and exits via the hypophysis, or pineal gland, radiating through the temples and the eyes. As mentioned above this constant in human biomagnetism resides in the spine and head, and applies to man and woman alike. Systematically, the front of the head (forehead) is positive, which is probably due to the energetic flow of the telluric force streaming outward at the level of the eyes via the pineal gland, and the cosmic frequencies coming in through the top of the head to access the pituitary gland before spreading down through the body. The back of the head is negative.

The cosmic *yang* energy enters through the top of the skull, I believe via the pituitary gland, and spirals down through the body exiting through the right foot.

The figure above demonstrates this trace of the spirals according to their research, I question the systematic configuration for both male and female and tend towards the belief that it is an individual affair. The following is an excerpt from their book, *Ondes et radiations humaines*, written in 1932 and translated by myself from the French:

Man's Electro-Magnetic Dispersion Field

The dispersion field or overall vibratory system of any human, considered in its totality, includes the following manifestations:

1. A vortex departing the heart, rising in a spiral to the right, penetrating the frontal parietal suture, describing a descending circulation in a closed trajectory (in morphology it is known that descending circulation is the characteristic of vortices), moving to the left of the body's central line to return to the heart. We have given this vortex the name "individual vortex," because it vibrates in a given color range specific to each person; it is the essence of each individual, so to speak. When a person dies, this individual vortex persists whereas all other manifestations expire; we leave this scientific observation to the reader's philosophical contemplation.
2. Two vortices imprison the body; one rises from the feet to the torso, presenting the characteristics of the telluric force with

ascending revolving cyclones. The other descends from the head to the lower body, presenting the characteristics of the cosmic force. These two vortices are not always in perfect harmony. When the telluric force predominates, passion is to the fore and physiological life more intense. When the cosmic force is dominant, the temperament is more cerebral, tending toward the spiritual.

3. An egg-shape form envelops the body from the top of the head to the toes, encircling the shoulders; it has a positive electrical charge in the right upper and left lower sections of the body, whereas there is a negative electrical charge in the lower right and upper left parts of the body.

4. There are two cylinder-shape force fields one within the other, with opposing polarities, the external one being positive for men and negative for women and the inner cylinder, negative for men and positive for women. In some individuals both cylinders are of negative polarity, as is their temperament. These fields are especially strong in newborns.

5. Spectral bands and rays, detected with our colored syntonisers, form what we call the "human spectrum". Figuring out these bands was our first task in the study with dark rays indicating damaged organs. These are the same phenomena that Walter J.Kilner perceived using dicyanin screens which he then named "The Human Aura" and clairvoyants refer to as the aura.]

This "individual vortex" in paragraph 1 seemingly corresponds to the notion—if one can consider the idea of vortices being the source pattern of flow in the physical world as we know it—of a personal pattern, somewhat like the astrological notion of the conjunction of stars and planets at birth or conception, affecting us throughout our terrestrial lives. If this pattern has a specific frequency, which is a reasonable enough assumption, one can further infer that such frequencies have ranges, which can be defined as colors, especially when referring to human health analysis with the idea of determining an individual's makeup.

This notion of cosmo-telluric energies flowing into the human body is also common to the ancient Chinese system of Daoism, the *yin* energy coming from the earth and the *yang* from the skies. As is often the case especially in ancient traditions, careful attention must be paid to the context of the terminology—the same word might be used in different contexts with which we are not necessarily culturally familiar. This energy consists of a range of what we choose to categorize but which probably cannot be separated in nature; for example, conduction, electricity, gravity, heat, induction, levity, light, magnetism, and sound are scientific concepts that we also do not fully understand but that sound quite familiar.

There does not seem to be any basis for reading more than a systematic pattern at work in the flow of energy here, and there seems no doubt that all physical forms, whether animal (including human), mineral, or vegetable, have what we can call a "biomagnetic" field.

Modern environmental conditions have changed so dramatically in the past few decades, with widespread electromagnetic and nuclear radiation prevalent in almost all countries of the world, that the earth's already weakened magnetic field has become even weaker. The ceaseless extraction of oil and other minerals, and a weakening resistance in the ionosphere to incoming cosmic energy, has inevitably had an impact on our biomagnetic potential. Ignoring these conditions can have unfortunate consequences, such as rapid depletion of energy, slow recovery from illness, fatigue, and even sickness, so it's a good idea to take the necessary precautions if you want to avoid exhaustion, especially if you practice energy work.

ELECTROMAGNETIC FIELDS

The scientific consensus would have it that our world is an affair of frequency, a region populated by infrared radiation and electromagnetic fields (EMFs), which have far-reaching biological, chemical, physical, and environmental effects. What's more, with regard to humans, EMFs are influential and sometimes determinative in coordinating and integrating our sensory, nervous, and endocrine systems. In simpler terms, we live in an invisible frequency soup and we're not quite sure what does what, to whom, when, where, or how but there is growing evidence that there is some form of relationship at work that affects our world, although not necessarily in a comparable way at the various structural levels.

This soup might interest you, because you are one of the ingredients.

One aspect of the electromagnetic field that cannot be explained by conventional theory (thanks to Don Hotson for pointing out this scientific detail) is that it acts at a distance, a fact that is perfectly manifest in the numerous psi phenomena. Furthermore, EMFs operate instantaneously—just like thought.

Whenever something moves, it creates a flow of energy. The circulation of oxygen and blood that results when air is displaced while breathing, the solar wind blasting past our planet, the creation of water from oxygen and hydrogen under the surface of the earth are ready examples of this.

It is all very well to measure these causes, recording the fascinating details and feeling content with that knowledge, but what do we know of the process? Unfortunately, we are satisfied to accept the details once they are measured as somehow being the effect and to leave it at that. While they may be true, they don't make up the *whole* truth but only a very small part of it. We systematically ignore the forms and patterns assumed by energy in its movement. These flows, as we have just seen, are essential to life. Of course, this seems so obvious that it is hardly worth mentioning, but the fact remains that movement is the agent of change, and those forms are the patterns of life, from which we can learn a great deal if we simply make the effort to observe them. We ignore them at our peril, for if we really want to understand how we can work harmoniously with Nature, we need to learn her ways rather than imposing our ways on her, which is ultimately impossible even if we vainly believe otherwise.

It matters little what we call whatever it is that flows in, on, or with these movements; we use them on a regular basis, and we can even observe this energetic flow at work in the environment, affected and causing effect when ambient conditions change. This is the origin of climate, which is the driving force behind all the products of Nature.

The picture is one of a totally interconnected, interdependent, and interactive whole, endowed with Consciousness and animated by an energy that defies definition (Consciousness) despite its evident presence on all levels of life, making itself noticeable even in its own absence, as in areas affected by nuclear desolation.

What's more, and not so generally recognized, is that these flows can act as a carrier for other frequencies, generating even further interaction and/or force fields. The actions of these flows may (or may not) depend on the strength of the original flow, or on their interaction (or not) with other materials or flows. In other words, it's a total mess if you're trying to figure out how it works and attribute a specific value to a given factor. That's the Maya for you.

But what do you observe? The evidence is there that our entire biosphere is alive, thanks to a remarkably well-organized system not only in which we live but that lives in us. That same system enables the whole to maintain all, even if we behave abominably towards it, showing no consideration or understanding. The arrangement either has immense forbearance, or it can handle whatever we throw at it. Hopefully, something of both.

Science teaches us that, historically, these flows in the troposphere (where we live) have been arranged and governed by Nature along EM frequency patterns established over several billion years. Science, however, is not so eloquent about the danger to the health of both humans and the environment that can potentially be partly or wholly attributed to a rupture in these flows, for example circadian rhythms, electric and magnetic fields, protein and water cycles, and other factors essential for the natural continuation of the world as we know it. So, while there seems little doubt that these rhythms are being disturbed, no answers will be forthcoming unless we look more closely at the consequences of the enhanced levels of nuclear and EMF radiation that have recently been created. In all likelihood an answer can be found in Nature, but claiming that we need more time to research it is disingenuous.

A recent June 2014, abstract of a medical paper, "60 Hz Electric Field Changes the Membrane Potential during Burst Phase in Pancreatic β-Cells: In Silico Analysis," by G. F. Neves et al., states:

> The production, distribution and use of electricity can generate low frequency electric and magnetic fields (50–60Hz). Considering that some studies showed adverse effects on pancreatic β-cells exposed to these fields; the present study aimed to analyze the effects of 60Hz electric fields on membrane potential during the silent and burst phases in pancreatic β-cells using a mathematical model. Sinusoidal 60Hz electric fields with amplitude ranging from 0.5 to 4mV were applied on pancreatic β-cells model. The sinusoidal electric field changed burst duration,

inter-burst intervals (silent phase) and spike sizes. The parameters above presented dose-dependent response with the voltage amplitude applied. In conclusion, theoretical analyses showed that a 60Hz electric field with low amplitudes changes the membrane potential in pancreatic β-cells.

The medical annals are replete with such papers on subjects related to the pathological impact of our modern conveniences, but contributions mentioning the potential dangers are few because there is no one to fund the research—surely the national utilities companies won't encourage such inquiry! It is so much easier to bury one's head and ignore the bad news.

Not that I am such a Luddite as to suggest abolishing electricity, mobile phones, microwave ovens, and other conveniences, but a little truth as to their effects on Nature or even just on ordinary men and women would do no harm, or are we obliged to be complicit in the cover-up? My grain of salt comes in the form of the solutions that Nature has enabled me to discover.

CHAPTER 5
MAGNETISM IN STONE

MAGNETIC POLARITY IN STONE

No geology lesson here, just a different take on the common or garden rock. No metaphysics, nor new age symbolism to be found here either. Stones have been around for ever apparently, several billion years we are told. We use, or disabuse them, much as we use the earth's resources, without any compunction or second thought, in – at best – partial awareness of their properties.

We know of their hardness, their weight, their density, their appearance, their strength, texture, porosity, resistance and so on. We apply those qualities carefully as a function of the intended use.

There is, however, a characteristic of stone common to all, varying only as a degree of intensity. I am speaking of a rarely, if at all, recognized phenomenon. Even among stone-masons, few if any, are aware of this.

A stone has a positive and negative polarity.

One on either side, or at either end. In the absence of another measuring instrument, take a pendulum in hand, suspend it a few centimetres over the rock. Keeping your mind free of any thought, without asking anything, the pendulum will start gyrating, clockwise for the positive magnetic polarity, anti-clockwise on the negative face.

Perhaps the most convincing way of witnessing this is when engraving a stone. A number of years ago, I went to see a laser engraver to seek his help in engraving some geometric patterns of my design on wood. Noticing that he was preoccupied on entering his shop, I asked if there was a problem. A big one, he told me, and showed me two samples of engraving that had just been completed by his laser technician. They could not understand why, even with the same settings on the laser machine for depth of cutting, intensity and time, the outcome was totally different between the two stones. One was very clearly defined and the image of the 9mm Eagle pistol leapt out of the rock; the other could only be seen with difficulty, there was no relief to the image, and would not be acceptable to the gun club making the order for fifty identical pieces. I asked if I could examine the two stones, and of course he agreed. The clear image was on the negative side of the rock, and the hard-to-see picture was on the positive magnetic side. I arranged all the rocks on the table negative side up which were then engraved. The client was delighted with the consistently clearcut image.

By way of demonstration, and for you to check your dowsing skills, here is a photo of a stone which you are invited to test to see which face, positive or negative, is facing upwards, and also to check its energetic value.

Perhaps try this yourself, find a round rock like the one above and take a chart with a scale of -60 to +120 inscribed on it; holding the chart in one hand over the positive side, and the pendulum in the other, ask what the score is for the stone. Having noted the figure, turn the stone over and repeat the same process. The score will be higher for the negative face of the rock. In other words, the negative face of a stone has a higher magnetic field of force than the positive side. If one is dealing with paramagnetic stone, one could expect to find a score of 5% higher or more than the positive side. Every stone, even if it is diamagnetic, will still show this positive/negative polarity, albeit with a lower strength than a paramagnetic rock.

The scale of the chart could represent the Centimetre-Gram-Second (CGS) system as proposed by Carl Gauss in 1832, later changed to a *gauss* as a measurement of magnetic induction. Basically, this is a measurement of how long one gram of substance, one centimetre from a magnet, takes in seconds to move to the magnet. In actual fact, the CGS scale is probably much lower in absolute values because magnetism is not an absolute, it is a function of the environment, a purely relative value yet a totally real and permanent influence in our troposphere.

The magnetic field on the surface of the earth ranges apparently from 0.25 to 0.65 gauss, as a function of the cosmic and telluric influences, not just the telluric ones as some would have us believe. There are most probably other influences also at work.

Having said that, please be aware that while this magnetic field is very weak compared to the field coming from an electromagnetic impulse (mobile telephony), electric power line, or even an electric cable in a house, it is powerful enough to influence the growth of plants and, of course, the state of health in humans. We are told that this effect is on a very subtle level; there are few studies to indicate, however, the range or degree, as a result we might best to conduct our own research if we plan on protecting ourselves.

We can reasonably assume that this influence has been present over the ages, unlike the electronic devices of the last sixty or seventy years; although they are of great consequence due to their intensity of power.

THE ELEMENTS

From what we can learn in ancient writings, and are readily able to imagine, the elements were subjects of great concern for our ancestors. Weather is still the most common subject of conversation throughout the world, in all cultures without exception – an indication of its importance to our well-being. Thunder was the subject of one of the earliest Indian poems in the Rg Veda (hymn 32 and 80 to the foremost deity Indra – associated with lightning, storms and thunder). Human concerns are very limited, consequently, we accept that as obvious and assume nothing more needs to be said or done. Alternatively, as we are doing here, we can attempt an understanding so as to improve our lot, as the ancients were so clearly doing. Fortunately the most complete legacy existing is to be found in Chinese texts, where we will now turn.

Climate, weather, planetary influences are the cosmic forces about which we can do nothing, save applying geo-engineering. We are subject to their unyielding power, however, if we are studious and observant enough we can learn to work with them, and perhaps even take shelter in time to avoid the incoming mischief.

It would be impossible to discover at what epoch this attitude all came about, and it is of no importance because it is such an essential part of the human make-up. What really counts is whether it is possible to make our immediate environment more comfortable, and if so, how. No matter the reason, if it is to hide from the extremes of climate, or to find a suitable source of food, or simply to feel good.

The Chinese cosmology, with the five elements as explained in their ancient texts, is full of symbolism and imagery; and not always easy for a western mind to grasp. With a little imagination and by juxtaposing phenomenal events with experienced reality, it soon falls into place.

These elements are best considered as four domains around a centre, to which the energies come from heaven. These energies are then manifest on earth, with an (official) in charge of each aspect. The earth (harmony) is the centre, where man resides, and the four seasons – wood (spring), fire (summer), metal (autumn), water (winter) assume their role therein. The dragon is in charge of water (of course water is underground, so is the dragon) and the Confucian virtue associated with water is wisdom (initiation).

The Five Elements so important to the Chinese way of being are not substances as such, but forms of energy. Better that one considers them as different manifestations of energy which inform every single substance and process of transformation, so not static but living, evolving entities, which of course they are. It is not easy for a person educated in the western (materialistic Greek) system to break free of the rigidity imposed by the accepted standard of physicality where life is only present in selected items.

The names of these elements are taken from observed physical phenomena, and the characteristics attributed to each of them are, of course, based on the qualities of the specific phenomenon, but originally and primarily, they should be considered as energy forms. When expressed this way with the aim of explaining a metaphysical, beyond the material, force with which we are all familiar, we can gradually perceive them as being the play of heaven and earth, and the fallout on us humans caught in between.

Of the two – heaven and earth – we are more concerned here by the earth, the telluric component for the simple reason that is where we can modify certain aspects.

TELLURIC FORCES

I have not seen it but it is said that there is a stele in China on which the ruler of that particular time, Kuang Yu (2205 - 2197 BC) had inscribed an edict on a stone stele which ordered the surveying of building sites to make sure those areas are not subject to underground water currents, so as to avoid the noxious influences, but not to get rid of them. I think one can reasonably say that no <u>ancient</u> site over two thousand years old, anywhere in the world, is ever found built on subterranean aquifers, unless those currents are incorporated into the overall design, as seen with the pagan precursors of the European cathedrals and churches of the middle ages, menhirs and dolmens of earlier times, because by doing that one can remove that harmful influence. The medieval builders reputedly even tested their building sites first by putting sheep there and observing if the sheep avoided the area. The Chinese *feng shui* experts, whom I have met and questioned, have no solution to resolve geopathic stress, for the simple reason, if you don't build over it, you don't have to remove it. In Europe today, only the Germans are concerned by this, even requiring by law for a *Baubiolog* (radiesthesist) to check the site out first, authorising the start of building.

But here is the rub, we in the modern world have little option but to build where space is available, and there is no leeway to position buildings out of harm's way – there is no space, so we have to make the best of a bad job. Having said that, the architects could readily pay more attention to integrative design principles and harmony rather than trying to excel by their 'originality' and eccentricity.

Telluric is the term used to define what comes from the earth – the Yin energies. These forces are inherent in the earth but sometimes they are found to be more than uncomfortable, especially when they derive from certain causes, such as geological formations, faults, underground streams, the crossing of waterways, underground caverns, mineral deposits, and such like.

What seems to be happening from the point of view of physics is that an individual geological formation can, and generally does, generate a movement which has its own very low voltage electric current. That current has a companion magnetic field, both of which of course interact with the surrounding environment, knocking on up to the surface where they are released from the constraints surrounding them below ground. Those forces apparently continue rising until such time that the incoming cosmic forces overcome and neutralise them, so to speak, which will happen at high altitude. Fundamentally, the movement and the path they take will always remain a mystery, we can only guess at the pattern as a function of the effects, but probably adding further to the complex of electric and magnetic fields of force.

It is extremely hard if not impossible to determine the precise nature and origin of a combined field of force, let alone its evolution, as not only is there a lack of available instrumentation outside of the laboratory to measure and gauge the energetic strength of that force, but it seems to be further modified by additional and highly random factors, familiar to Traditional Chinese Medicine (TCM) practitioners, such as time of day, atmospheric conditions of temperature and pressure, planetary influences, in addition to the subterranean determinants which are not categorised. Even then, there is a singular lack of instruments sufficiently refined to measure the fine degree of fluctuation in movement or its impact. A parenthesis while on the subject of instruments – Philip Callahan, one of the inspirations for my interest in the towers – used a device of his own creation, the Photonic Ionic Cloth Radio Amplifier, to measure ELF and VLF frequencies, as detected coming from the round towers. He patented this device in 1993. It is no longer available as originally conceived, except in the extremely limited form of an earth-sampling device.

This is where the experienced radiesthesist, versed in the subject matter of frequencies and their possible interactions, equipped with the necessary charts and suitable tools, and sufficiently practised to ask questions of the most diverse kind, comes into use.

Whilst these frequencies may be tiny, representing a minute force, that does not mean they are innocuous and without effect. For it is bound to have an effect even if that is not immediately apparent. On what, to what extent and how are the rub, but it IS going to have an effect. Hence the need for us to assume a much broader approach than the current cause-effect, force vs. power, materialistic view generally shared by humanity.

I would further venture that our modern conveniences, electricity, radio, telecommunications and transport, as well as the use of synthetics in clothing and footwear, have altered that weakened capacity, and we adapt to the extent that we bear it all but with disharmony as a companion. As we know no different, we are perhaps spared greater suffering.

Such are the possible characteristics of the telluric forces that are at work round the clock. When they adversely affect humans and animals, which often occurs, they contribute to, and form one aspect of what we commonly refer to as geopathic stress (GS).

This magnetic component is especially relevant to us earth-dwellers because those frequencies coming from the ground rise up to the surface where we hang out, and can work serious mischief.

It is nigh on impossible to determine all the unitary effects at work in our environment. Let alone which ones are impacting us, because not only are our metabolisms different, but our magnetic fields are too, changing with our state of health, diet and subjectivity to planetary influence.... There are so many potential factors at play that it is positively daunting trying to calculate what is going on.

These geological events have their own footprint, so to speak, when the energy they produce reaches the surface of the earth and interacts with whatever is present there. As these events evolve, it naturally becomes quite difficult to appreciate and assess their effect because we do not recognize their impact, once again lacking the instruments to measure their effect on the human immune system or even the organs, and even less, do we record statistics in their regard. What is more, there is a constant interaction of varying degree as these energy forms move and interact. As an example, the energy field developed using the system of wooden rods that I use to offset the telluric influences causing geopathic stress often travels extensive distances – four to five kilometers – until another stronger energy source is encountered and weakens, or the contrary, the first. No rules here.

When general health problems (which can, of course, assume mental, emotional, or physiological dimensions) arise, however, we are keen to find causes, and if these energy forms are part of the equation, which indeed they are, it might be easier not only to explain the source but, far more importantly, find a solution.

Low-field magnetic influences, so neglected by just about every branch of modern 'science', have so much to account for. Going back in time and to China, we find not only the earliest recognition of the dangers of underground water in the form of *feng shui*, because of what it does to people, but what is far more pertinent, the movement inside the body of energy as it flows (or stagnates) in the meridians, as recognised in traditional Chinese medicine.

Of the most common telluric events, one finds:

- Underground water—in any form—flowing through any underground passage, capillary, pipe, fracture, or fissure. The resultant magnetic field can be surprisingly high in frequency and strong in effect. The field fluctuates depending on multiple factors, including what the water is conveying, the speed at which it flows, and/or whether it interacts with any other earth energy. The earth's natural energy field is particularly influenced by the crossing of two or more watercourses, or by the crossing of a watercourse and another type of energy line.

On a scale of worst to good, the crossing of two underground waterways, irrespective of their individual depths, is the most dangerous phenomenon for humans with regard to biological damage, due to the intensity of the combined magnetic influences on reaching the surface.

- Geological formations, with their different mineral and magnetic components, are responsible for a lot of mischief, but mainly as a function of their ability to allow the flow or directional thrust of water; deposits of magnetically or gas charged rock are also to be contended with.

- Geological faults, caused by movement in the tectonic plates, or the result of the upheavals the earth has experienced, also affect the energy field at the surface. While they are less common, mainly because they do not often present zones suitable for construction, they can be very harmful.

- Underground caverns cause problems for the inhabitants above due to the resonance developed by the substances collecting or passing through the cavern zone. Invisible from the earth's surface, these holes are constantly creating energetic movement due to pressure, interacting with surrounding formations of a chemical and/or physical nature, and the natural passage of the effluents upward to the zone of least resistance—where we reside.

Basically, these energies when creating frequencies that interact with the environment and the people who are in the vicinity at their point of exit, represent a potential threat. Obviously, but not necessarily so, those people with a weakened metabolism will be affected more than robust individuals. They will, however, further weaken the immune system as the frequencies are simply not conducive to a harmonious restoration at night when mitosis takes place.

Like humans, most large animals, cows, horses and dogs prefer to live in the absence of those telluric influences. On the contrary, ants and crawling insects are happy there – as are cats. Perhaps that is why the cat was sacred to the ancient Egyptians, for they seem to have an unusual ability to handle geopathic stress quite well. Do they transform it to something beneficial? What is the story here?

The multiple categories of telluric forces as explained by many dowsers, although of interest in the way the earth possibly functions, are of no practical use in my opinion. They are one more anthropocentric demonstration of 'knowledge' because nothing of benefit can be derived for humanity, let alone Nature, even if you know theoretically that it is a "vile vortex", "Curry/Hartmann line" or some such supposedly at the root of the problem.

If you, as it would appear our ancestors were, are of a practical rather than a hypothetical bent of nature, and are more interested in achieving concrete results it is probably best to leave these investigations to the 'theoreticians' and focus on what can be really accomplished, irrespective of the fanciful, scientific names people give to phenomena that no one really understands, let alone can prove, because they are essential components of Nature.

Remember that telluric forces are natural but constantly evolving influences; regrettably, we have little real understanding of them and perhaps even less concern as they stream into our environment, which is increasingly cluttered with vibrations emanating from power grids, electrical circuits, computers, mobile phones, and routers, amplified by metallic structures and furniture, all of which play havoc with our vibrational balance, leaving Nature's creation to struggle in their midst with very little help coming from the cosmic energy that in the past balanced the whole. They will not go away just because we ignore them.

Since being here on earth, we humans have had sufficient time to learn and become familiar with Nature's ways. Now while that is a very long time, surely much longer than the average history curriculum would have us believe, there is very little, if any, information which has worked its way into the practical stream of knowledge. This is true with regard to academic knowledge, but there is every reason to believe that our forebears knew a lot about the energetic influences coming from the ground. The reason for saying that is because they left proof of the remedies they found to deal with certain issues.

Especially the ones that caused problems.

GEOPATHIC STRESS (GS)

It would appear, however, that ancient **peoples such as the Celts and Egyptians, realizing the potential utility of these** subterranean **forces**, integrated them into their vestiges and practices. Early modern humans in all probability used, among other devices, large standing stones—menhirs and dolmens—to offset the effects of these influences, or alternatively they steered clear of them and built their homes elsewhere. This type of geometric and natural "science" was probably transmitted at a later date by various orders—Templars, Cistercians, guild-members—because the effects of these forces are demonstrably active even today despite the major modifications made in recent times to many European cathedrals and monuments built during the Middle Ages.

The problem is that such telluric influences can and do cause deformation on a cellular level after a certain time – a completely unknown factor in the equation. These can be dangerous, even mortal for humans.

One of the main factors in maintaining health is another neglected, and basically unknown one, namely geopathic stress (GS). The first thing to do on discovering the presence of GS, is to remove it. For the simple reason, a person who exists in the influence of that condition, cannot get better so long as they stay there. On the contrary, there is every chance their health will deteriorate.

The geopathic stress caused by underground water can be fatal; it is almost always detrimental to health, and any understanding as to its nature has been lost over the ages to the extent, we are not even aware of it, let alone know that something can be done to resolve it.

The ancients (Sumerians or Babylonians rather, Chinese, Egyptians, Greeks [normal they applied what they learned at school in Egypt] and Indians) knew not to build on zones that are affected by what is below ground and works its way up to wreak havoc on the surface. We know that fact by inference alone, because no ancient monument is built over such problem zones, but that is a strong argument when attention is paid to that important fact.

Unawareness, ignorance and greed make for poor bedfellows as we well know, however a very large majority of the urban population pay with their health as a result.

There is no need to call upon any so-called authoritative science in my mind, for the simple reason, there is none that studies the effect of geology on the inhabitants of the troposphere, or that would be able to either explain or remedy what is going on.

It is of interest, however, that the medical profession has since around 1900 recognized GS as a pertinent factor. A medical doctor in Germany, Gustav Freiherr von Pohl, found that a number of his cancer patients all came from the same village. When he asked a dowser to go and check, the latter found that all the patients were subject to GS in the form of an underground flow of water over which all of their homes were located along one side of the street, and in addition they were all sleeping in the zone of influence.

Since that time, there are a few doctors, mostly involved in the alternative fields, paying attention to this phenomenon, but far too few to make the necessary impact of awareness on the general public. The damage caused can only increase as a result of unawareness and irresponsible construction. At least in Germany nowadays, one is required to call on the services of a dowser before building living quarters.

Any builder with a few years of experience under their belt will be aware of the effects of certain terrains, yet probably be quite unable to explain the causes of geological phenomena. Subsidence is one of the more easily explained problems. The capillary action of subterranean water can only be determined on a micro-scale with radiesthesia; the excellent quality of the British hydrological survey maps around the world are most useful but on a large scale. Rising damp, poor drainage and fungal issues can be happily, albeit expensively, resolved. Solutions are there for a restricted number of problems. GS can cause cracking in concrete or walls, often along the line of flow of the unseen influence, frequently leading to insect invasion or infestation – they love the low-energy areas.

We are, by and large, totally unaware of the forms and patterns assumed by energy in its movement. The exceptions, of course, being found in aerodynamics and certain disciplines of the physic sciences. These flows are inherent and essential to life, even if we do not appreciate that. Movement is the agent of change.

Those forces are the patterns of life, from which we can learn a great deal if we simply make the effort to observe, and even better, cautiously experiment with them. Conversely, we ignore them at our peril, for if we really want to understand how we can fit harmoniously into Nature, we need to learn Nature's ways rather than attempting to impose our ways on Nature.

People, animals, plants and objects regularly spending time in zones subject to geopathic stress are likely to experience problems at some stage in their existence. *Per se* GS does not cause illness, but it can definitely lower immunity and is constantly interfering with and altering the frequencies we depend upon for our material existence. Having said that, the strength or otherwise of the individual's immune system is all-important, to the extent that some people sleeping over GS never experience difficulty.

GS is a component cause, because of the frequential disturbance, possibly making things worse but definitely not reducing aggravation. For most forms of life and its normal development on earth, a balance in natural energetic vibration is required, or for the human and animal, resonance in the nerve and endocrine systems. So, for every living cell, whether human, animal, plant or mineral to grow, develop, mature, age and finally die, from old age and wear rather than from disease, it must, mandatorily, vibrate harmoniously.

The influences of geopathic stress depend on the source type or origin which are numerous, both natural and man-made; such as geology, geomagnetism, electricity, electromagnetism, cosmic forces, shape and form, meteorology, their combination and periodicity, to name just a few.

The effects on humans can be transferred and/or transported, slow or the contrary and they are generally cumulative over time. Even the World Health Organisation, if it can be taken seriously, reckons that 30% of buildings have what is known as Sick Building Syndrome, or structures subject to geopathic stress. I would put that closer to 60% from experience these last twenty years.

This is further complicated (for humans at least) by factors such as metals (beds and chairs), chemicals, poisons & pollutants, malnutrition/diet, water quality, recreational drugs, alcohol, psycho-social stresses, genetic weakness, electrosmog, chemtrails and so on.

In nature, some plants and animals are happy (within limits) in geopathic stressed areas, while others are not, and suffer. Earth-bound lemurs, entities or ghosts are generally happy there too.

The nature of such problems is varied, but appears to cover a vast, if not the entire physical, emotional and mental range: e.g. addiction, ADHD, aggressiveness, allergies, anorexia, anxiety, arthritis, asthma, bed-wetting, bulimia, cancer, candidiasis, cell de-regulation and growth issues, chronic fatigue, cot death (SIDS), dementia, depression, diabetes, eczema, enzyme production, exhaustion, foetal development, food intolerance, glandular fever, headache, heart disease, hyperactivity, infertility, insomnia, intestinal disorders, lethargy, lymph problems, memory loss, miscarriage, multiple sclerosis, myalgic encephalomyelitis (ME), nervous disorders, obsession, pain, Parkinson's, premature birth, resistance to treatment, restless sleep, rheumatism, schizophrenia, sexual abuse, skin problems, stressed relationships in partnerships, stroke, tuberculosis, etc., etc..

The effects do not only concern health: houses with geopathic stress running through them are consistently slower to sell than those without (it would appear that people sense something's wrong even if they are not consciously aware of it), lightning strikes such areas on a regular basis. GS is often a factor in struggling businesses, accident black spots, failing technology, corruption, financial decay and bad luck in all its forms.

Initially for humans, GS often manifests as a sleep disorder which is perhaps explained by the fact that mitosis, cell replication, occurs during sleep. If a person sleeps in a GS-free zone, the cells develop in harmony with the original state of the cell. However, on the contrary, if the ambient surrounding frequency stops the cell from vibrating at the desired level, the cell becomes deformed in relation to the original. Imbalance and disharmony follow.

Bed and work are where most people spend the majority of their time. While it is sometimes sufficient to move the sleeping area or desk a few feet and the danger is avoided, that may not always be possible. Placing a copper or aluminium sheet under the mattress will reflect the radiating waves from water, for example when an underground stream passes below, but that will have a contrary effect if there is a geological fault, and its effect would be accentuated. The best solution is invariably to use a harmonizing system as it can further usefully offset the effects of the other components that complicate our lives: radio masts, high voltage power-lines, industrial processes, domestic electric appliances and the numerous sources of daily exposure to magnetic and electromagnetic fields.

They say that GS appears in diverse forms, such as lines, beams or combined pattern forms. That is quite possible and even probable; it even varies according to the time of day and night, as well as atmospheric conditions. Such awareness seems quite academic to me, and of little practical use. I would like to, and do, provide solutions, not more fascinating information concerning the increase in technology of the last few decades and its devastating effect on our natural environment.

Detection of GS is easy on humans. Using a pendulum, simply place the pendulum over the Triple Warmer 2 point (Gate of Humours). If there is no GS, the pendulum will indicate a positive response, if there is GS, it indicates a negative response. Detection of the telluric influences from the ground can also be revealed by an adequate radiesthesist, although very often the sensitive human will feel a variety of sensations as a function of the nature of that force.

To my knowledge no scientific biomedical journals have published any articles in the English language concerning Geopathic Stress. Having said that, it would appear that some doctors take the notion seriously, so there is a glimmer of hope that some suffering may be curtailed, but there is no doubt whatsoever, if we find that it is an issue best to deal with it ourselves as well as we can.

If we are really sincere in wanting to understand how we can fit harmoniously into Nature, we have every interest in observing and emulating Nature's ways rather than attempting to impose our ways on Nature. Such a way of thinking would presumably have been axiomatic in the minds of people who had survived disasters or were living in memory of such cataclysms that destroyed the dinosaurs, others that provoked the ice age, and so on.

The upheavals caused by the flood of the earth and the events that preceded it have, in all likelihood, left us with a very different living space from the one which existed beforehand. No matter if you believe that series of events or maintain that the earth is a more recent affair, the fact still remains, geologically and tellurically speaking, the place is currently complex, and we do not have a science worthy of the name to aid us navigate.

If there was a branch of learning that studied the effect of geology on the inhabitants of the troposphere, we would know that certain subterranean conditions are dangerous, and ignored at our peril. There is no such branch of official learning today, but that does not mean it never existed, as indeed it did - and still does, but as radiesthesia or dowsing, (defined as pseudoscientific by Wikipedia, the renowned authority on subtle energy, as they are on pyramids, the famous tombs of pharaohs, even though it still remains for a body to be found in one).

It is especially relevant to note that most, if not all, of the medieval cathedrals built in the 12th and 13th century, were located above these powerful subterranean water passages, the architects seemed to have known that these sites were connected, so to speak, by these underground waterways, and the construction sites were chosen as a consequence of that awareness. Although that awareness antedated the cathedrals, which in many, if not most, cases were built on older, pagan sites. Once again, as we will infer from other examples later in the text, authorship of available knowledge is usurped and closeted by the powers-that-be.

Having realized the danger of these subterranean forces, and found ways to detect them, it is but a step to surprise a way to do something about them.

And there can be no doubt that the ancients did, for they integrated those remedial measures into the vestiges which we still have among us. It is from these that we are able to discover that they not only knew something which we no longer understand, or has been occulted, but they were knowingly applying those methods.

As previously mentioned the magnetic component of stone is closely related to water, largely due to the distinct intelligence of water. When the two are combined to form a mutually supportive magnetic field, it seems that a lasting coherence can be created, resulting in a harmonious environment that can carry some distance, as a function of the telluric influences that will interfere, but only when reaching the surface, so that benefit can be conveyed several hundred metres, or even several kilometres.

As also stated earlier, the most harmful telluric influence is the crossing of underground water, while it may be a joy for the wyvern, it is very deleterious for man. Systematically, standing stones are to be found erected on the exit point of those currents. It is no coincidence that the pillars of the medieval cathedrals are also erected on those points, as is very often the case with the standing stones, obelisks (when in their original positions), and ALL the authentic round towers of Ireland.

One can presume the reason for placing the stones, as we saw earlier with the experiment Schauberger made on the disappearing spring, was to encourage the spiralling pulsation and levity of water from the depths of the earth, drawing the wyvern up, so to speak? The advantage of course is that the effect also modifies the danger of those magnetic fields, and in addition can create a space of intense harmony and peace for human, animal and plant. What is more in that zone, the energetic influence is sometimes sufficient to change the frequency over a much larger area.

Our forebears well understood that poor soil means unhealthy plants and sick consumers. Crop-destroying bugs are Nature's scavengers, keeping the place tidy, so if you want health for one and all, the trick is to take care of the soil. If you don't have a ready supply of volcanic (paramagnetic) dust, use compost, correlate which weeds bring the minerals up to the surface, develop mycorrhiza thanks to worm population in the earth, and of course, magnetic force, the natural frequencies absorbed and retained in rock.

The use by post-diluvian humans of stone would appear to offer a key insight into the understanding of what the ancients knew about the magnetic force of stone. Their palette of skills ranged from the use of large standing stones, obelisks, structured stone-laying, to geometric form, among other devices, used to offset or palliate the effects of these influences. It is still possible for us to perceive these, but it does require a certain flexibility of intellect and openness of mind.

This type of natural knowledge was probably closely guarded by the fraternities (the girls surely had better things to do, although were probably members of the original gangs too!), guilds and groups. I would venture that such protective reasoning was not out of an interest in any form of control of society, but more a question of maintaining precision in the analysis and application. Alternatively, with a recent apocalypse in living memory, water and mud floods, which destroyed most of the world some few thousand years ago, the survivors, unable to make head or tail of the Egyptian hieroglyphs, decided a written communication was not the way to go. Obviously, the last sentence presumes that the Egyptians were antecedent, that they had something to write about, and that it was worth transmitting to future inhabitants. There are other examples from around the world where we have lost the coding cipher, maybe even with the crop circles too, we are unable to perceive the message and just have to learn the way of hard experience with massive loss and destruction, in the hope that we can learn in time to prepare for the next catastrophe!

I digress. Naturally, at some stage such esoteric know-how is viewed as magic, and given that public opinion is easily corrupted, especially when vested religious interests are in the balance, there were probably some protective measures taken, such as not writing these notions down and committing them to memory – of the select few, as apparently in the Druidic tradition.

It might be pure speculation, but the story of the Templars that follows emphasizes this human tendency to strive for the upper hand when financial profit or control offers some benefit, or how things become perverted in the process as soon as greed is involved, and the greater (for Nature at large) benefit is soon forgotten.

Given the evidence the ancients left behind in the form of standing stones of various sorts, and there are a huge amount in the world, despite so much having been destroyed or deliberately removed, it is astonishing that we have not figured out what is going on.

THE COSMIC ASPECT

So far I have placed emphasis on the telluric interaction of this magnetic effect of stone – its interaction with the earth due to the water current below ground, and the evacuation of that energy to the skies where it seems to be 'dealt with' by the incoming cosmic force. There are, however, other ways of considering the possible interaction of the forces involved.

The 'antenna' approach, as proposed by Philip Callahan, in his book, *Ancient Mysteries, Modern Visions*, discusses that the round towers may have been designed, constructed and used as antenna - huge resonant systems for collecting and storing wavelengths of magnetic and electromagnetic energy coming from the skies. Based on his studies of the forms of insect antenna and their capacity to resonate to micrometer-long electromagnetic waves, Professor Callahan suggests that the Irish round towers (and similarly shaped religious structures throughout the ancient world) were human-made antenna which collected subtle magnetic radiation from the sun and passed it on to monks meditating in the tower and plants growing around the tower's base.

He reckons the round towers were able to function in this way because of their form and also because of their materials of construction. Of the sixty-five towers, he states that twenty-five were built of limestone, thirteen of iron-rich, red sandstone, and the rest of basalt, clay slate or granite - all minerals which have paramagnetic properties and can thus act as magnetic antenna and energy conductors. Callahan further states that the mysterious fact of various towers being filled with rubble for portions of their interiors was not random but rather may have been a method of "tuning" the tower antenna so that it more precisely resonated with various cosmic frequencies.

Callahan also postulates that the geographical arrangement of the round towers throughout the Irish countryside mirrors the positions of the stars in the northern sky during the time of winter solstice. Apart from the difficulty for earthbound humans to find any access to such a layout, there seems little, if any, practical use. Of all his theories concerning the *raison d'être* of the towers, this seems to be the least probable. As a radar technician responsible for the guidance of aircraft, he would most likely have been in possession of a copy of Rude's Star Finder and Identifier, published at the Hydrographic Office, Washington DC in March 1942 under the authority of the Secretary of the Navy, a very useful device for a plane pilot/navigator, or other, to plot one's location from the position of the stars and planets in the sky. In Appendix there is a copy of the northern celestial equator as found in the Rude Star Finder. For those interested, it could be an idea to compare that picture with the frontispiece of George Lennox Barrow's *The Round Towers of Ireland*, and so get an idea of what he proposed.

Quite what such a comparison adds to the discussion I fail to grasp as it seems to be a sensational suggestion that the ancients felt a need to lay out a map of the stars on the ground in Ireland, or have I missed the point? Who would use such a map, for what purpose, how could it be used? The questions are multiple, the answers are vague at best, and it seems to serve no practical purpose whatsoever.

As to his idea of the towers being antennae, once again a number of simple questions relegate the idea to the same basket of 'interesting', impossible-to-prove and, practically speaking, rather useless information when applied to the possible use of incoming frequencies in the absence of a method to benefit from such an arrangement.

The beauty of putting these ideas out there into the aether-space of people's minds is that there is a strong chance that somewhere down the line a small piece of the puzzle will be added to the whole.

The judgement above may appear harsh, particularly on someone who can no longer defend themselves, especially when and what is more important, they rendered tremendous service to the subject. However, I stand by what I say as the situation today far exceeds any niceties we should politely extend to each other, for I am convinced that we are in serious need of remedies to resolve the sorry state Nature is in due to our abuse.

That takes us to the next stage of the investigation, but before we go there it might be a good idea to contemplate the possible motives for erecting the round towers in Ireland.

The erection of stones above ground in geometric patterns whether in groups or standing alone is also generally easy for us to appreciate because we know from the agricultural standpoint how useful it is to know the time of year, from the astronomical viewpoint how instructive to predict and prepare, from the ritualistic angle how beneficial for the local society. Imagine how those substantial benefits would be surpassed if the added value of well-being were to be included.

Consequently, assuming that there is a functional benefit for the setting up of a stone structure, if and once that benefit can be proven, subsequent to experiment, to be the result of a specific means or method, there is a sound basis for presuming that those means are relevant in accomplishing that benefit, and that its employment is what the builders intended by its application. That is the approach adopted in my research, which must then be considered, discussed, experimented and explained – where possible. Practicality is of the essence, what works to our benefit without adversely impacting Nature at large. All other hypothetical and subjective considerations, as proposed by various proponents are left to the individual belief system, if and as you find necessary.

Such a way of dealing with the subject matter immediately narrows the scope. By the term practicality, I imply necessity of the most simple kind, the need for survival of the individual, group or community, resulting in well-being and continuity. It seems that without those two components, the recipe for society soon disappears as a people become disgruntled and move away, or at worst, die out.

The difference being today that we choose to ignore that interfering with planetary and telluric frequencies is fraught with consequence. It is only when we respect Nature and remain within the bounds of reason that we can hope to maintain existence on an even keel. There is every indication that former mega-settlements were destroyed, perhaps as a result of man's folly, which encourages me to persist in this obscure research.

There are, however, several substantial shortcomings, as I see it, which could benefit from a human effort to restore our lost ability to commune with Nature in a more meaningful manner, and in so doing, achieve a greater harmony for all concerned.

As you can intimate from above, a familiarity with magnetism is key. Another aspect which deserves greater focus to acquire a more complete view of the whole is water. The relationship between water and stone could be the subject of a substantial volume of literature but must be left for another day.

You will perhaps have appreciated that underground water is a potentially lethal component in the mix we are considering, it can however be turned to our benefit, and that is what the ancients appeared to be doing. The instances of underground water, especially the crossing of water currents, as a cause for serious problems are so numerous amongst radiesthesists, shamans and dowsers there is little room for coincidence in this respect.

Of interest and possibly of some importance is the magnetic layout of the stones used in the construction of the Great Pyramid of Giza. All the stones appear to be with the negative magnetic polarity at the top. And the whole edifice is over the crossing of underground water. If the demonstration of this understanding and application of magnetism and water is still to be found in Egypt, there is a further avenue of possible explanation for that knowledge reaching Ireland with the arrival of the Egyptian Scota with her Greek husband, Gaythelos, and their company. But yet another rabbit-hole!

PART II
CONNECTING THE DOTS

CHAPTER 6
THE ROUND TOWERS

**What is not proved and what does not exist are
the same; it is not a defect of the law, but of
proof.**

It is almost axiomatic to say that the more an opinion is
categorically and repeatedly expressed, the greater attention
should be paid to the underlying motive, for it is increasingly
apparent that wishful thinking is, or could well be at the base in this
specific instant.

The history of Ireland from the fifth century onwards is so closely
tied to the Christian church, it has now become an impossibility to
discover the facts of what transpired before the time of the early
Christians arriving, and there would be little exaggeration to say
that the popes considered all people as pawns in the power games
they were wont to play with anyone who had any clout, which they
might be in need of. In view of the fact that Ireland was "given" to
the English king, Henry II in 1155 by pope Adrian, the history – if
one can call it such – from thenceforward is centred on ecclesiastic
affairs and those who applied the will of the strongest, to the natural
exclusion of any other than the victor's version of the "truth", a
common enough phenomenon in our ongoing epoch of
psychopathic dictators. So little interest is given to the wellbeing of
the inhabitants and their manner of living in such a situation that

only the wielders of power are given space in the books, solely recording their benefits and feats accomplished, if any. And so it would seem that towers fell into that category of convenient oblivion as concerns their origin.

The round towers for those not familiar are the subject of a mystery which, while it has not perhaps attracted as many theories, or as much ink to flow as the Egyptian pyramids, are still one of those unresolved, intriguing phenomena, because not only are we still very much in the dark as to their purpose, who built them or when, but why only in Ireland, Scotland (three of them), and supposedly in Douglas, Isle of Man.

They are such singular structures confounding in quick succession the reasons put forward by modern writers and theorists for their being there in the Irish countryside – and some in urban areas (obviously of more recent development). Sometimes they are to be found in large open spaces with a dominating view for anyone perched in the top, other times in the depths of a valley; there is no apparent logic to be found there. However, given that two facts have never, to my knowledge, been considered either with regard to the round towers, nor in environmental studies, risk casting some light onto the enigma of the towers that would justify a reappraisal of the mainstream paradigm, and more importantly, make way for the further necessary research into their *raison d'être*.

There have over the course of time perhaps been a hundred or more towers according to resources worthy of credibility, but in the general consensus now maintained, only sixty-three remain. In Keane's book, mentioned below, he speaks on page 305 that lists of round towers amount to one hundred and twenty, with the remains of sixty-six or so remaining. Some of that number are dubious by my restrictive criteria (crossing of underground water and magnetic layout) but that changes little when taking a larger view of the purpose which could well have been to use stone for its magnetic properties, its benefit for us and the immediate environment.

The dimensions of the towers vary little which would suggest the work of a coherent body, such as a guild working in full understanding of what they were doing. The skill and finesse deployed in the construction of the towers is worthy of folk belief in the person of the Goban Saor, or more likely, his lineage. It is unlikely that one hundred towers or so could be built in the lifespan of one man, but that is mere conjecture. If the criteria of modern guilds are anything to go by, one could project those standards backwards to the predecessors, namely secrets retained by the guilds are only passed on to worthy students, those who have demonstrated their ability over numerous years, otherwise they tend to be lost. Even today, the building trade rarely demonstrates the same expertise shown by the craftspeople of the guilds, perhaps it was similar then, albeit to a lesser extent, as there was less demand for construction, but the principle remains, hold on to the good stuff and make sure it is passed on down to posterity.

Not enough is known about Gobán Saor, Ireland's mythical builder, for any constructive factual evidence to be put forward. No doubt clues are to be found in the mythology, even if those stories are interwoven and cloaked in Christian myth with a strong affiliation to the legendary Saint Patrick and other missionaries.

It is a mere hypothesis to suggest that the origin of the towers was concomitant with the foundation of the monasteries. It cannot be proven for lack of dated epigraphic, literary or any other form of evidence, and serves no chronological purpose whatsoever.

We should be aware, nevertheless, of the fact that although the church has never actually, by means of any form of statement, encyclical or other, claimed to have built the towers. It is far more subtle and pervasive than that, by insinuation, it has allowed passing generations believe it is at the origin. A clever method reinforced by the gradual building of ecclesiastical buildings beside the towers, and in some cases onto the towers, as in Turlough for example.

Far more sinister and malevolent, however, because it not only places a firm proprietary grasp over the towers and their environment, but it distorts the beneficial effect and energetic impact on the immediately available agricultural land, the church has buried numerous bodies all around, up to the walls of the towers even. A serious pollution of the immediate area of many of the towers, even if Barrow would have us believe that towers were sometimes built over existing graves. His bones would appear to have an extra long survival time in the humid Irish soil.

Barrow claims that "there is no doubt that the use of lime mortar, like the principle of the arch, was unknown in Ireland before the coming of Christianity in the fifth century." That, like the horns on a hare, may well be a possibility, but debatable on the scant evidence for or against. Why Christianity was responsible for any architectural development in Ireland at that time, however, remains quite nebulous, especially if there is such clear sharing of skills in metalwork and crafts amongst the Celtic tribes of western Europe – not that Ireland was necessarily inhabited by Celtic tribes. Knowledge of those techniques would probably be due to active exchange and travel amongst people who had common interests. The early Christian settlements were primitive structures, sufficient to provide shelter, and the concept of permanent buildings came later, when the wherewithals to provide material and labour were forthcoming, but not until then. Were Christians the only travellers to Ireland in those times?

There is no point being drawn into any polemic, so let's stay with what we know and can find with the towers, then as now, for that is where the truth lies, the why of the towers, not the who and how of the monasteries, churches or stone structures added on a later date for their perverse aims.

However, the technique used in building the towers is NEVER used in any of the ecclesiastical buildings, reputedly associated with the former. What is more, the Roman Church is silent on the application of this technique, as it is concerning the technology used in building the medieval cathedrals, so there is a strong case for the builders being totally dissociated one from the other.

Let's have a closer look at the towers themselves and their chief characteristics.

DESIGN CRITERIA

In Barrow's book, he usefully records of what type of stone the towers are comprised, not all of those left standing but a total of fifty one towers. Quite a few of them are made with a mixture of stones, for example limestone for the main body work, whereas the window and doorway frames of granite. That information is not only of academic interest, it should also encourage us to get our thinking caps on. The three towers (which meet the underground water crossing criterion) in Scotland, Abernethy, Brechin and Egilsay are made of sandstone too.

Just down the road from where I live on the Sheep's Head is a fine stone carver, who has given me insights into the work and tribulations of the modern stone-worker, as well as information concerning the ins-and-outs of the work itself regarding the hardness of stone and the necessary tools when working on the various categories of stone. I went to see him to ask if he could give me a rough idea how long it would take to dress a stone to be used in the building of a tower. His reply was that a ready answer would be impossible given that the hardness of the stone would basically determine the time spent, but the tools available (speaking from the standpoint of those currently accessible) would also be a key factor, as well as local working facilities – water, shelter (you don't work so well huddled for hours over the stone you're working on if the wind is driving the rain onto your body). One cannot help but wonder if the builders of the towers did not have access to more sophisticated tools than we are led to believe were available in the early days. We can justly wonder what kind of tools did those masons have to work the granite and sandstone? You cannot dress granite with a copper chisel, even with the best will in the world! And sandstone is no picnic either.

Of the fifty-one towers in Barrow's book only five are made of slate, the easiest of stones to work; two are made of granite and thirteen of sandstone, one of basalt and the remaining thirty of limestone, which next to slate is relatively easy to work. Sandstone is almost as hard as granite and requires tough tools to make an impression on it.

Most of the stone used to build the towers came from local sources apparently, but that is the easy part to quarry and transport the stone. To dress a block to the approximate curve of a fifteen metre circumference requires a lot of effort, skill and time. This creates another enigma, much in line with the question concerning the tools available to the Egyptian builders of the pyramids, but perhaps we have enough to deal with already without introducing another enticing rabbit-hole.

The circumference of the towers measures, by and large, between 14 and 17 metres; there are two walls of block stone and mortar – one external and one internal filled in with rock rubble and earth; the thickness of the two walls at the lowest point at which it can be measured ranges from 0.9 to 1.7 metres. We assume this from the stumps that remain. Doorways, windows/openings, the height of storeys and diameters are also on clearly defined patterns. The elevation of the doorways varies substantially, from ground level, as at Scattery, to almost 8 metres high at Kilmacduagh.

A consideration that needs to be mentioned is the fact that none of the round towers still have their original cap. At some stage they have all been replaced, destroyed or fallen, so we are not even sure what form they took, and the reports on the repairs effected in the late 1800s remain silent as to the original state, if ever it was known. The few caps that have been found nearby, as with Aghagower, are conical in form with a slant of approximately 65°, but a complete one has never been found. Consequently, it would not be unreasonable to question if a tower did have a cap, and if so what might the actual form of it have been.

The reason for questioning this is important because the state of 'cap' or 'no cap' potentially makes a substantial difference to the energetic flow through the body of the tower, and the subsequent overall effect on the environment.

These shared particularities would indicate applications and concerns totally at odds with the mainstream theories of the towers destined to be belfries or for purposes of sanctuary.

How, for instance, are you going to get a bell through a doorway five metres above ground?

If your bolt-hole is 8 metres above ground, you are going to need a ladder of some kind to climb up there. A ladder that long will not bend as it is pulled up through a "doorway" measuring only 1.6 metres high. If the ecclesiastics opted for a rope ladder, how do you get the ladder down from the doorway when it is needed? I know I would hesitate using one left constantly exposed to the elements, resting on damp stone, where it will rot rapidly. Your sense for practical concerns ladies and gentlemen is somewhat dim.

It is not only improbable that security could be the purpose, but highly unlikely. It is a sound argument to postulate that the height of the doorway ensures stability as the structure will not be weakened by an opening, but that is simply side-stepping the issue – at which a certain expertise is now apparent. There must have been a reason for an opening at different heights when there are so many other standard features common to all towers.

Stone was used in Ireland from earliest times. In passing, it is not possible to provide precise dates as modern methods, such as Carbon-14 testing, rarely used by geologists as it can only date an object younger than fifty thousand years or so, might be able to tell you the age of the stone, but not when the structure was built. Stone in Ireland was used as a building material for a number of purposes such as fortifications, tombs apparently, beehive huts and simpler structures – not many of which remain, but it was only, in all probability, from the time that mortar (cement) was introduced that higher, sturdy buildings such as the towers became possible. There are of course many examples of stone being used beforehand, but not in a context of 'building'. Stones have been laid out and positioned in many different forms, and therefore one can reasonably assume, with different purposes. But nowhere else outside of the areas of Celtic influence in Ireland and Scotland can you find (still standing) construction of a 30 metre tower with a double wall and openings that defy all practical use of a conventional opening, let alone of a building with recognised purpose.

There are a number of very evident details that would point to the construction of the towers being the work of a guild or some form of esoteric brotherhood, somewhat like the builders of the medieval Gothic cathedrals, given the water and polarity principles discussed here, and the skills required to determine those. There are multiple facets common to all towers, apart from those principles: dressed stones in most of the towers – consider how long it takes to work and shape a stone, then calculate how many stones are used in the construction; the height and orientation of the doors and openings cannot be haphazard; the difficulty in capping the tower precludes that being the work of an amateur.

If, as so many theorists have suggested, the church was at the origin of the towers, a few very simple questions should find ready answers, but they are met with resounding silence:-

1. Why is the round tower part of an ecclesiastic enclosure, and what is the purpose?
2. Why is the polarity principle present in the tower not to be found in <u>any</u> of the other ecclesiastic buildings in the same site?
3. What is the purpose of the polarity principle?
4. Why are the towers positioned over the crossing of underground water?
5. Why are graves dug in close proximity to, and around the towers?
6. Why are the doorways at different heights? And how do you withdraw a ladder into the building, especially one 8 metres tall such as the Kilmacduagh tower?
7. Why has a bell never been found in a tower, or the infrastructure to support one?
8. If as suggested, the towers were built in the 6th/7th century, is that not three hundred years before bronze bells were cast?
9. How is it possible for a hand bell chime to be heard from 30 metres up?
10. How do you fit a bronze bell through the narrow doorway and lift it up to the top?
11. Why has the polarity principle never been used anywhere else if it is an ecclesiastic technique?

Quite when and under what impulse the towers came into being is pure conjecture, albeit with the generally accepted arrival of lime mortar when building became significantly easier rather than relying on a dry-stone structure, although the dating of that is uncertain. Even if a time for the arrival of lime mortar in Ireland and provenance were known, the fact still remains, the church never used the technique which provides well-being, and by not recognising its very existence declares itself forfeit both for its original use, and its authorship.

The round towers in Ireland have been in turn a subject of active discussion, a bone of contention, and of quaint but virulent Victorian dispute before falling into the very large pot of archaeological question marks. What is intriguing is the persistence of the interest they infuse over time, with periodically a book appearing on the shelves – even if the book has nothing new to add. It is as if there is an urge to stress the official dogma that it was the church that built the towers, for that is the inevitable line spun, with absolutely no evidence to back up that supposition.

My aim is to add to that lengthening list of books, but with the difference of the addition of an original theory, which can not only be demonstrated, but actively experimented – repeatedly, so more than just a theory.

There would be little point aligning the varied positions, theories and conclusions presented in the last few hundred years with regard to the round towers in the light of what I present here, no one has ever mentioned the polarity or the water-crossing principle in their respect, so it is of use perhaps to stress some of the other aspects which other authors propose.

It is not original as regards the science behind the technique, there is a very strong chance that this application has been around for ever in the neglected history of humankind, but due to the reasoning as to how and why they were built. The *who* and *when* pale to the point of insignificance once there is an indication as to the *why*. Nevertheless, the *why* would preclude certain parties from the construction work due to their inability, their ignorance of the skill or art, thus clearing the decks of a certain number of firmly held beliefs.

Ireland is unique in that not only have so many stones been left in place – despite the horrendous destruction of the last century and a half, but there is a tradition so firmly attached to the land and the people on it that it creates a unique backdrop on which new arrivals – whether ideas or peoples – are absorbed by the tolerant atmosphere prevalent in the land. I dare say this is common to other ancient lands (I think specifically of India) where a tradition has been continuously maintained over the ages, despite the cruelty of invaders, colonisers and imposed regimes, but the low density of the population here in Ireland and the huge number of ancient sites still accessible and apparent is rare to find. There is also another traditional factor at play in Ireland, respect for the belief, no knowledge, of afterlife therefore ancestral reverence. If you plan on staying alive and healthy, you would not cut down certain trees, nor move, let alone remove certain stones. There is a very strong affinity with the natural world, particularly trees, stone and water which the invaders never managed to obliterate.

It is as if there is an unspoken principle practised by religions around the world – when they do not have the strong arm of their national or mercenary armies to do their dirty deeds. If you cannot enforce, then adapt and in a few generations you will succeed if you maintain the pressure. This is very apparent in Ireland where the church usurped the Celtic/Druidic tradition, not by crushing ancient belief and imposing the new religion, but an inexorable, pervasive pressure applied on all aspects of the traditional life. The result is what we see after sixteen hundred years of domination, even though that authority has been severely shaken and brought into question in recent years as a result of the revelation of the base human behaviour of people who preached otherwise. It is not inconceivable that some mitred heads lose their sleep wondering if we shall ever wake to the fact that the violence wrought is not just with the sword, but by the word.

If it were really the case that the Roman church was the source of these extraordinary principles used in the construction of the Gothic cathedrals, or the round towers of Ireland, there would or should be documentary evidence in the Vatican archives, and how the money to embark on the round tower building project, let alone the cathedral building campaigns of the 12th and 13th centuries, was found. What is listed as being publicly available in the vast Vatican Apostolic library (https://www.vaticanlibrary.va/) makes no mention of any such material, and all of my enquiries produced no useful result. Of course, that site does not include texts possibly located in the Vatican archives.

There is reason to believe that the Knights Templar treasure trove, at least the literary component with the architectural know-how, went down with the ship.

As William Stirling in his 1897 book, *The Canon*, said "The priests were practically the masters of the world… freemasons, or some body corresponding to the mediaeval freemasons, with exclusive privileges and secrets required for building the temples under ecclesiastical authority, have always existed. And the knowledge we possess of the mediaeval freemasons is sufficient to show that their secrets were the secrets of religion, that is of mediaeval Christianity."

This can only be partially true, they were the secrets held by the Templars, but not theirs by any means, otherwise they might not have come to such a sticky end. If the church were at the origin of what I refer to as the Geographic and Polarity Principles used in gothic architecture, they would have used them in subsequent construction, but that was never the case, and the pope sided with Philippe le Bel due to political expediency rather than any intellectual conviction in removing the Templars.

If the Templars had perhaps revealed that knowledge and its provenance, rather than keeping it under wraps, it would not have been the subject of such jealousy. No, the secrets 'found' by the Templars dated from a previous age of which we shall probably never learn more.

As can be seen in Rosslyn Chapel, a great deal of symbolism is revealed by the early masons who built that remarkable structure, but not the Geographic Principle, so it would appear that the freemasons were not privy to the secret, even at a period so close to the disappearance of the Templars.

CHAPTER 7
MAINSTREAM BUT DUBIOUS PROSE

Here is a brief overview of some of the 'literature' regarding the background to this farfetched claim that the church was at the origin of the towers that has been repeated incessantly over the last two hundred years – yes, that is sufficient, for it is now fixed in the collective mind.

In 1832, at the instigation of George Petrie himself, recently elevated to be a member of the council of the Royal Irish Academy, it was resolved to hold an essay competition on **the origin and uses of the round towers of Ireland**. In 1833 the prize of a gold medal and fifty pounds was awarded to Petrie himself for a **20-page document**. It is perplexing why neither Petrie's 20-page document, nor his 558-page book are available on the website of the RIA, whereas his watercolours are.

I have selected three texts of relevance to the round towers – not for what they reveal, but as a demonstration of the indoctrination portrayed – published between 1833 and 1999 presented and/or discussed here: George Petrie in 1833 (but only published in 1845), Barrow in 1979, and Lalor in 1999. These three, and many other texts from celebrated academics and antiquarians, persist in setting forth the establishment position of the unproven Christian origin. There is a resounding absence of convincing evidence, overt wishful thinking and denigration of those with other views as seen in Petrie's opening paragraph below.

A non posse ad non esse sequitur argumentum necessario negative, licit non affirmative. Hobarts Reports, 1641. If a thing be not possible, an argument in the negative may be deduced, namely that it has no existence; but an argument in the affirmative cannot be deduced, namely, that if a thing is possible it is in existence.

GEORGE PETRIE (1790-1866)

Here is what George Petrie had to say in his **558 page** book entitled *The Ecclesiastical Architecture of Ireland anterior to the Anglo-Norman Invasion; comprising an essay on the Origin and Uses of the Round Towers of Ireland,* (there was promise of a second volume but it never appeared despite the fact that the author survived another thirty years):

Here is what Petrie proclaims:

"The question of the origin and uses of the round towers of Ireland has so frequently occupied the attention of distinguished modern antiquaries, without any decisive result, that it is now generally considered beyond the reach of conclusive investigation; and any further attempt to remove the mystery connected with it may, perhaps, be looked upon as hopeless and presumptuous. If, however, it be considered that most of those inquirers, however distinguished for general ability or learning, have been but imperfectly qualified for this undertaking, from the want of the peculiar attainments which the subject required — inasmuch as they possessed but little accurate skill in the science (if it may be so called) of architectural antiquities, but slight knowledge of our ancient annals and ecclesiastical records, and, above all, no extensive acquaintance with the architectural peculiarities observable in the Towers, and other ancient Irish buildings – it will not appear extraordinary that they should have failed in arriving at satisfactory conclusions, while, at the same time, the truth might be within the reach of discovery by a better directed course of inquiry and more diligent research.

Hitherto, indeed, we have had little on the subject but speculation, and that not unfrequently of a visionary kind, and growing out of a mistaken and unphilosophical zeal in support of the claims of our country to an early civilization; and even the truth – which most certainly has been partially seen by the more sober-minded investigators – having been advocated only hypothetically, has failed to be established, from the absence of that evidence which facts alone could supply.

Such at least appears to have been the conclusion at which the Royal Irish Academy arrived, when, in offering a valuable premium for any essay that would decide this long-disputed question, they prescribed, as one of the conditions, that the monuments to be treated of should be carefully examined, and their characteristic details described and delineated.

In the following inquiry, therefore, I have strictly adhered to the condition thus prescribed by the Academy. The Towers have been all subjected to a careful examination, and their peculiarities accurately noticed; while our ancient records, and every other probable source of information, have been searched for such facts or notices as might contribute to throw light upon their history. I have even gone further: I have examined, for the purpose of comparison with the Towers, not only all the vestiges of early Christian architecture remaining in Ireland, but also those of monuments of known or probable Pagan origin. The results, I trust, will be found satisfactory, and will suffice to establish, beyond all reasonable doubt, the following conclusions:

I.   That the Towers are of Christian and ecclesiastical origin,
        and were erected at various periods between the fifth and
        thirteenth centuries.
II.   That they were designed to answer, at least, a twofold use,
        namely, to serve as belfries, and as keeps, or places of
        strength, in which the sacred utensils, books, relics, and
        other valuables were deposited, and into which the
        ecclesiastics, to whom they belonged, could retire for
        security in cases of sudden predatory attack.
III.   That they were probably also used, when occasion required,
        as beacons, and watch-towers.

These conclusions, which have been already advocated *separately*
by many distinguished antiquaries — among whom are Molyneux,
Ledwich, Pinkerton, Sir Walter Scott, Montmorency, Brewer, and
Otway — will be proved by the following evidences:

For the first conclusion, namely, that the Towers are of Christian
origin :

1. The Towers are never found unconnected with ancient
    ecclesiastical foundations.
2. The architectural styles exhibit no features or peculiarities not
    equally found in the original churches with which they are
    locally connected, when such remain.
3. On several of them Christian emblems are observable, and
    others display in the details a style of architecture universally
    acknowledged to be of Christian origin.
4. They possess, invariably, architectural features not found in any
    buildings in Ireland ascertained to be of Pagan times.

For the second conclusion, namely, that they were intended to serve
the double purpose of belfries, and keeps, or castles, for the uses
already specified:

1. Their architectural construction, as will appear, eminently
    favours this conclusion.

2. A variety of passages, extracted from our annals and other authentic documents, will prove that they were constantly applied to both these purposes.

For the third conclusion, namely, that they may have also been occasionally used as beacons, and watch-towers:

1. There are some historical evidences which render such a hypothesis extremely probable.
2. The necessity which must have existed in early Christian times for such beacons, and watch-towers, and the perfect fitness of the Round Towers to answer such purposes, will strongly support this conclusion.

These conclusions — or, at least, such of them as presume the Towers to have had a Christian origin, and to have served the purpose of a belfry — will be further corroborated by the uniform and concurrent tradition of the country, and, above all, by authentic evidences, which shall be adduced, relative to the erection of several of the Towers, with the names and eras of their founders.

Previously, however, to entering on this investigation, it will be conformable with custom, and probably expected, that I should take a summary review of the various theories of received authority from which I find myself compelled to dissent, and of the evidences and arguments by which it has been attempted to support them. If each of these theories had not its class of adherents I would gladly avoid trespassing on the reader's time by such a formal examination; for the theory which I have proposed must destroy the value of all those from which it substantially differs, or be itself unsatisfactory. I shall endeavour, however, to be as concise as possible, noticing only those evidences, or arguments, that seem worthy of serious consideration, from the respectability of their advocates and the importance which has been attached to them.

These theories, which have had reference both to the origin and uses of the Towers, have been as follows:

FIRST, as respects their origin:

1. That they were erected by the Danes.
2. That they were of Phoenician origin.
Secondly, as respects their uses:
1. That they were fire-temples.
2. That they were used as places from which to proclaim the Druidical festivals.
3. That they were gnomons, or astronomical observatories.
4. That they were phallic emblems, or Buddhist temples.
5. That they were anchorite towers, or stylite columns.
6. That they were penitential prisons.
7. That they were belfries.
8. That they were keeps, or monastic castles.
9. That they were beacons and watch-towers.

It will be observed, that I dissent from the last three theories, only as far as regards the appropriation of the Towers exclusively to any one of the purposes thus assigned to them."

----

12 years after winning the first prize for his meagre 20-page contribution, Petrie publishes a book where from pages 5 to 122, he refutes the theories mentioned above, and then from page 123 onwards expands on the features of Christian architecture and decoration in Ireland and elsewhere, with absolutely no application to the round towers, let alone any mention of their origin and use. Only as of page 358 does he expound on belfries when he offers more obscure arguments in support of his preferred theory. It could be usefully recalled at this point that a belfry was designed to house a bell, generally made of iron coated with copper. Although the Romans apparently used bells in London to announce the time, the

CHRISTOPHER FREELAND PH. D.

belfries were probably in open but populated areas, which does not fit in with the locations of many, if not most of the towers in Ireland. Built in the late 11th century in France and Flanders especially, bells were destined for urban areas where local communities could be reminded of church services, time of day, emergencies or summons. The larger bells often required several men to ring them, they were so heavy. Few and far between are the round towers located in urban areas, especially of that period, and even fewer with a doorway large enough to allow the passage of a heavy bell that would then have to be hoisted up the remaining twenty-odd metres.

One can legitimately wonder why for more than a hundred pages Petrie deals with the designs on capitals and bases of columns, as they are totally absent from a round tower, but he insists on inundating us with his architectural expertise, while regrettably adding precisely nothing to the matter in hand.

Given his constant reference to the findings, measurements and deductions of other researchers, it appears that Petrie relies on the observations of the same and he perhaps spent little time inside a tower, especially in view of so many anomalies in his book:

- He never mentions the double wall construction employed in all authentic towers.
- He maintains that the protuberant base of the Clondalkin tower is original, and he even compares it to a tower, with a similar base according to a 17th century seal, that was blown down in the 13th century, and a bulbous construction in Wales.
- He claims the towers were topped by a stone cross – there has yet to be found one.
- He states that as no pagan building in Ireland resembling a round tower has ever been found that only Christian builders had the necessary skill.
- He further states that prior to the arrival of Christianity the Irish did not have the know-how to make lime mortar, nor make an

arch despite "innumerable" remains of buildings. He omits to list the remains and in the absence of proof of such, it makes for a shaky foundation to lay such a broad claim.

Although he frequently admits that the ecclesiastic building(s) in the vicinity of the tower were built several centuries after the tower, he never addresses the why or wherefore.

George Lennox Barrow (1921-1989)

Apart from the intense work Barrow did over a period of eight years or so on his book *The Round Towers of Ireland*, he added nothing new to the paradigm – in fact he persisted with the same story that the church was at the origin of the towers, despite stating: "Documentary sources on the towers are few and unreliable." He built a very useful gazetteer or compendium of towers actually in existence, which had reputedly existed or were referred to in the annals, with photos, drawings, measurements and local lore. His references are uniquely drawn from sources that maintain the mainstream paradigm, ignoring or ridiculing any texts that challenge that. For example, he dismisses peremptorily the work of Henry O'Brien, with "this farrago of romantic and mystical nonsense." Not exactly the type of open mind one might reasonably hope to find in the 20th century with regard to the only other person to take part in the Royal Irish Academy competition of 1833, winning second place, £20 and a bronze medal for his 509-page document. Especially as O'Brien was only 25 years old giving his all against a man of 53 who managed a rather paltry 20 pages.

In this book, there is something very disturbing about the entry for Kilmacduagh, which was restored in 1878-9 by the Board of Works. The entire complex of ecclesiastical buildings was cleared and revamped, including the graveyard, work was done on the tower, supposedly, to install a lightning conductor. No mention is made of the state of the tower by Barrow, however he cites extensively from an undated letter to the Editor of the Irish Builder, from an anonymous J.A.F., where one learns on page 29 that the tower was fast decaying like the other buildings; that the tower was completely restored, as "a great portion of the south side had fallen; and there were evidences to justify the gravest apprehensions regarding a considerable portion of what remained." It is explained in the letter that the interior of the tower was excavated in order to install the conductor, which led to the discovery of human remains (four bodies in all apparently) below the level of the foundations. The archives of the Irish Builder can be found on: https://archive.org/details/irishbuilderengi2118unse/page/29/mode/1 up?view= theater.

Now while we don't know how far down, to what elevation of the tower, the barrel was reduced, Dunraven's photo shows the east-facing window on level 5 to be intact. From there to the earthen floor is a distance of almost sixteen metres, and from the earth floor to the soft soil where the skeletons were found is 5. 80 metres. So almost 6 metres of earth had to be dug and raised 7.62 metres to remove it from the inside of the base, with the last two metres of depth being dug out from a roughly circular hole in the stone foundations that measured 1.54 metres on average. Clearly not large enough for anybody but the smallest of people, and four of them at that, and some of the skeletons protruding out beyond the stone foundations.

If this tower was built in the ninth century – two hundred years after the earliest building date estimates, that means it took a thousand-odd years for the soggy, acidic soil to **NOT** break down the skeletons of four pagans. A man of the cloth would not have gone digging around in a graveyard to build a tower is Petrie's argument. Either the pagans were made of sterner stuff than us, or there is another option, such as a rapid disposal of genocide victims from the 1850s era at the foot of a tower in a graveyard.

Predicting the time for a set of bones to revert to the earth is not a precise science as it depends on the acidity of the soil, humidity, the condition of the body, and so on.

Ei incumbit probatio qui dicit, non qui negat. The proof lies on him who affirms, not upon him who denies.

BRIAN LALOR (1941-)

The second in the dogmatic, must-be-church series is *Ireland's Round Towers – Origins and Architecture Explored*, by Brian Lalor in 1999.

On the very first page, one reads an opening statement: "Round Towers are the only form of architecture unique to Ireland. ....an informative guide to these intriguing products of the Celtic imagination."

It makes it hard to have any confidence in what Mr Lalor might have to say thereafter if he is not aware of the towers in Scotland, which he obviously is because on page 96 of the same book he devotes half a page to the existence of the Scottish towers and the one in the Isle of Man, but offers the strange statement "...none of these buildings can seriously be considered as significant influences on Irish early medieval tower building. None retains a conical cap, and as all appear late in terms of the door typology, cannot be considered as forerunners to the *cloicteach*, but as parallel developments or evidence of the influence of Irish building practice..."

Firstly, it is pure conjecture to say the towers are medieval, but more significantly, even Barrow in his book states "The only towers with complete caps... are Clondalkin and Devenish, and only the former <u>appears</u> to be in its original state." (Underlining is mine).

Because the tower does not have a conical cap it is dismissed despite the fact that it is a well-established fact that few if any of the towers have their original caps!! They have been reworked, perhaps restored in some rare cases, but definitely modified in the last few centuries. We have absolutely no idea what the original form might have been.

The "products of the Celtic imagination" would imply that Mr Lalor knows something that no one else is privy to. Generally, common sense and as of recent times, when an out-of-place object is found anywhere, the scientific approach is to search, to cast the net far and wide so as to bring as much light as possible onto the enigma. A shame. Incidentally, if anything is out-of-place in this context, it is the ecclesiastical buildings in proximity, sometimes even attached.

When one reads this type of prose, one comes away with the very strong impression that new ideas are clearly frowned upon, even ridiculed. That is a very manifest sign that something needs to be concealed, the familiar "Nothing to be seen here, move on!"

The last two authors never mention the Pelasgic, Pelasgian or Cyclopian architecture that is often employed in the towers. This technique is found in the ancient world, Egypt, Greece and South America as can be seen from the photos in Appendix 3. It was Henry O'Neill in his brief but fact-filled 56-page document, The *Round Towers of Ireland* dating from 1877 who brings this to our attention, especially as concerns the Lusk round tower.

What can one expect from the church establishment? The very word speaks a volume on the epithet. Something established does not move. We humans love such organizations, they make us feel secure and enable us to feel that everything is in its place. Somewhat like God's garden – an appealing figment of the imagination depending on order, structure, discipline, but most importantly – belief.

The belief that such an establishment structure is the most suitable solution to organize the material world is a very common one, and why not. There is probably no alternative in actual fact, however free-thinking we may like to believe ourselves to be!

Problems arise when the hierarchy is upset, for example when a closely-held belief is questioned. This would apply to all domains of the establishment – medicine, academia, religion, the press and publishing, for there is no separation among the component elements, they are mutually dependent in their need to maintain their hold over society. This is merely an observation, not a judgement, for there is absolutely nothing I can do about it, even if I felt so inclined.

The truth, however, has a habit of coming to the surface, even if no one is looking when it gets there. But if there is a deliberate effort to make it sink again, to remove it from public view by obfuscation, ridicule or bulldozing, we can reasonably assume – due to past experience – that there is something authentic going on and our attention should be piqued.

Without going to the extreme of saying that the establishment has made every effort to conceal, denigrate and deride Irish culture in general, it would be quite fair to say that certain aspects of Irish history have been eclipsed in order to diminish the contribution played by the inhabitants of that island – irrespective of their tribal appurtenance, creed or politics.

The church (the Roman version) took a long time to work its way to the top of the establishment pile, and as is systematically the case when a group of men gather (especially when the girls have been evicted), their egos encounter no restraint in achieving their short-sighted aims, and they are obligated to step on their original human values, crushing all resistance by whatever means at their disposal. If one can piece together the fabric of the Irish form of Christianity, scattered as it was and no match for the Roman (church) bulldozer, much like their legions of a former epoch, it appears to be a more humane version. The religious schools of Ireland were all the rage in western Europe for several centuries, with students coming from far and near to learn, and their alumni travelling throughout Europe becoming saints and renowned teachers.

There is one thing that is sure, and even the establishment experts agree, there is NO documentary evidence to support that it was the Roman church which was at the origin of the round towers.

The third author to have written a book on the round towers is Alanna Moore. Despite her useful input as concerns the practical aspects of model towers made of sandpaper, along the lines of Callahan's suggestion, she has a tendency to fawn to the established opinion, confounding her belief with what the authors above have to say. Often even adding rather obscure statements, as with 'there would have been a monastery neighbouring those towers found in isolation', without providing any support for the statement; or 'danu is the word in Sanskrit for streams of water', not sure where that one came from, but definitely not Monier-Williams Sanskrit dictionary.

According to her, there are 75 towers in Ireland despite the fact that she frequently refers to Brian Lalor's work which lists only 73, including vanished towers! She finds Barrow historically inaccurate because he dated the towers to pre-Christian times. And even stranger, she states that George Petrie precedes Henry O'Brien despite the fact that Petrie, as a member of the Royal Irish Academy first called for papers on the origin and purpose of the Round Towers in 1833, to which O'Brien responded. It is not very factually convincing when she further states her belief that the Atlanteans may have been an advanced amphibious race. Why not? But quite how that fits in with the subject supposedly at hand, I fail to grasp, and even less how she can claim that Brian Lalor wrote 'the most comprehensive tome on Round Towers to date'.

Once more, so much new age myth is asserted in her book *Stone Age Farming*, it makes for a very confusing swamp of information, techniques and fable which cloud an already murky horizon. One might reasonably expect an "expert" dowser to make a clear statement as to what might be below the ground of each and every round tower, however that is not to be and one is met with a hash of info impossible to confirm or invalidate, serving little save generating further questions. The beauty of dowsing is the possibility to find answers, definite yes or no responses to precise questions.

So, if we could start anew and look at what we still have, and contrary to what has been written to date, look at the <u>why</u>, rather than the when or by whom, we stand a better chance of learning what the practical applications of these remarkable structures were, and far more importantly, **ARE**.

CHAPTER 8
KEY FACTORS RELATING TO
THE ROUND TOWERS

"An island in the ocean over against Gaul, to the north, and not inferior in size to Sicily, the soil of which is so fruitful that they mow there twice in the year,"
- Diodorus Siculus, ~60 B.C.

Until such time that we have a better, let alone complete, understanding of water, we are doomed to a stop-gap strategy regarding water management, and probably as a consequence, all aspects of agriculture. A lot can be deduced from the unconventional and unorthodox ideas of Viktor Schauberger on water and its management thanks to his in-depth understanding of its functions. He would seem to be one of the only men in recent history, he started writing about his observations of Nature in 1929 or so, to have expressed his concern for the superficial, and ignorant way we treat water. The foremost idea discovered by Schauberger was the role of water temperature in its health and he stressed the importance of the 4°C transition temperature. Followed by a series of bright minds that added further key facts, mention can usefully be made of Chaplin, Coats, Korotkov, Mae-Wan Ho, Pollack and Voeikov should you wish to delve deeper.

What is most relevant in my understanding, however, is what is going on below ground. If we want a more complete picture of the story of water, we need to take a look at what happens to it during the invisible – because underground, stage where it seems to spend a substantial share of time in its life-cycle. For that is from there the round towers are drawing their influence.

The fall in groundwater, or the water table is perhaps the main reason for failing crops, lacking the necessary mineral nutrients which are brought to the surface with the flow of water from below. Rainwater only moistens the earth, there are practically no nutrients absorbed by the soil when it rains.

Thales of Miletus (around 614 BC) was on to something when he said that water is the only true element, as all the others are derived from it. Another Greek, Plato, states in his *Timaeus and Critias* that the people of Atlantis regulated water with the help of warm and cold water. Water needs a balanced, organised and regulated temperature system to develop.

The natural cycle as regards the build-up and decomposition of vegetation involves a layering of the substances in that vegetation one on top of the next, the principle is the same as one finds in the making of compost. Due to the effects of temperature and pressure, because we are underground with the substantially different constraints found there, those substances are changed into carbon compounds with the help of water and local conditions. It would be more than probable that the water is also decomposed in this process, resulting in the production of more gases that will release further carbon compounds, such as carbon dioxide and/or carbonic acid, which then start their journey upto the surface through the layers of the vegetation above, causing further movement and transformation in the chemical compounds in their passage. Given that the air mixture down there in the bowels of the earth is quite different from the conditions found in our biotope, and perhaps

127

even absent, totally new conditions and possibly compounds are developed, leaving us pretty much in the dark as to the movements and transformations underway.

We can only deduce what that alchemy is from the end-products that reach us, which will then be further modified thanks to the help of the kinetic force from the magnetically aligned stones in the tower drawing them upwards. It seems clear that the enclosed environment below ground is the source of new forms of vegetation and substances such as ore and rock, adding to the potentially magnetic mix with its influence on water and the surrounding substances. This is all in line with Nature's governance, ruling temperature, pressure and their exchanges, giving a coherent and sequential picture, which if only we observed more closely, we might get a batter handle on the processes involved. Those intricate procedures at the varying depths, with the changes in temperature and pressure, subject to the internal movements all go to form natural laws which we can deduce if we are attentive. There is a strong chance that in past times – the contempories of the builders of the towers, might well have understood. I suspect that Schauberger was also privy to such thanks to his observation.

What exactly is needed for growth? Where does the right balance lie? Obviously, Nature holds the key, so every effort made to respect what we observe and profit from in this cycle. For it is thanks to that sequence of functional events in the correct sequence and conjunction that refinement of the substances in question are accomplished, in turn leading to further avatars in due form and time. This all happens very much thanks to the special movement (spiralling perhaps) of water, as it oscillates between the kinetic and potential so generating the electromagnetic current with its accompanying magnetic phenomenon, which of course are not static and evolve in their progress, interacting as move on their merry way.

Nowhere else can one find such a satisfactory explanation that naturally leads to many more questions, but fundamentally expresses the mechanics and probable physics that were appreciated by those beings who built the round towers. There can be no coincidence for there being a number of details systematically found in common among all authentic towers which would logically preclude any question of haphazard coincidence in the deliberate and consistent application of those aspects. Given that,

All round towers in Ireland are located exactly above the crossing point of underground waterways.

Every single stone is laid with the negative magnetic polarity facing upwards.

Every tower is made with two walls, one external and one internal, reinforcing the draw capacity.

Why would you pay for a mason to dress every stone of a thirty metre high building, not with just one wall, but very often the two walls? And then to stack the stones in a very specific order?

Dressing stone, for those not familiar with a mason's work, is a tedious task at the best of times, complicated by the hardness of the stone. It involves cutting the stone to the curve which the roundness of the final structure determines, and requires a certain skill if one is to respect the rules of the art. The towers taper as they rise, so the diameter of the walls recedes as one gets higher, requiring the stones to be increasingly rounder on their external face. And this was often accomplished both on the inside and outside walls. That represents an awful lot of work and dedication. We should not forget either that a mason's work is generally performed outside, exposed to the elements at all times, even if a rudimentary shelter was erected to provide some comfort. A harsh and taxing trade which leaves most men in poor health at a relatively early age.

A very simple instance is never mentioned when dealing with ancient construction work – the cost. Not that the expense is of any great consequence to the whole question, because we do not know what the benefits were most of the time, and we never shall in all probability, so it is not a relevant consideration until one ponders on the reason for going to such pains. One might reasonably assume that the benefit was worth all the effort. When that effort is reproduced elsewhere as in the case of one hundred or so towers that once existed in Ireland, there must have been a substantial advantage for all that work, otherwise there would be little point in doing so.

There is little purpose discussing the thrust of all the numerous arguments put forward on the subject, as I have already said, but if an open mind could be brought to bear on the points raised here, I believe it would be a small but sure step towards learning more about the peculiar and unique energy generated by these buildings. Perhaps that might lead to a better understanding of their objective, which I would here suggest had something to do with well-being for the local community and the environment.

It is difficult, sometimes impossible, nowadays to access the towers, for reasons of security and inaccessibility for a host of motives. Even if one was able to do so, measuring sealed doorways, climbing upto heights without special equipment to verify orientation and measurements, let alone the possible damage one might cause, would serve no purpose, and would not make us any the wiser.

However, if we take a leaf out of Callahan's book *Ancient Mysteries, Modern Visions* where he mentions the benefits of strong paramagnetic fields on agriculture, supported by ideas expressed by Davis and Rawls in *Magnetism and its Effects on the Living System*, as well as Dunlop and Schmidt in *Biological Effects of Magnetic Fields,* we can start to comprehend the possible aim the ancients had in mind when they stacked a pile of stones, thirty odd metres high, in the same magnetic alignment. You are going to obtain something very powerful thanks to the intense magnetic field of force created.

What is that something?

This is the opportune moment for me to stress that what I say and practise derives from a sincere effort to work in harmony with Nature, given what I perceive as an imbalance that seems to have been created due to human greed and lack of consideration for our phenomenal world, and I strive for practical, rather than fanciful or theoretical solutions which might serve to re-position men and women in our natural god-given environment. I would venture, on the strength of what I know from my experiments with stone, plant and animal (including us men and women) life, investigation of old traditions from around the world, and an understanding of life sciences, that such a strong magnetic force is a way to boost the movement of the life force, so enhancing – perhaps even encouraging as it is alive – the component elements of that mysterious essence.

It would appear to be an accepted physical fact that two electromagnetic fields imposed on each other at 90 degrees is a classic scalar formation, the result being a carrier wave. Incidentally for the radiesthesists reading this, the Universal Pendulum indicates that the frequency coming from a round tower is white in the magnetic phase.

So that carrier wave, an analogy for which would be the electric phase which carries the current to the light bulb, coming from the energies generated by the electromagnetic fields of the water currents below combining as they rise to the surface of the earth is retained within the stone structure, which seemingly directs that energy in the alignment of the negative magnetic polarity until such time that it is released through the doorway and windows of the tower. When combining the magnetic field of force from the polarity-aligned stones with the carrier wave coming from the crossing of underground streams, the resulting effect is what would appear to be the high energy that the ancients recognized as what Nature needs. Given that a pendulum can indicate the direction of flow of water, it is normal that one asks when examining a tower, in which directions are the flows of crossing water going. It is from there that one can deduce that the direction the doorway is oriented is determined as a function of the flow downstream of the main/larger water current.

The questions come thick and fast, but there is little chance of answering them so let's go back to the energetic benefit for the land as revealed by the comment of the fisherman ferrying Callahan out to Devenish island that the grass was finer on the island than on the mainland.

In his book, *Paramagnetism* on page 70, Callahan recorded measurements with his specially designed device at the Glandalough tower doorway and found a reading of 20mV. Quite what the significance of such a score is he does not say.

If we share no reference points with what the ancients were doing it is quite normal, when you think about it, that our science cannot correspond to what we might be looking at here, there is a chasm which cannot be bridged for the simple reason we are lacking all means to relate our modern world with what went down then all those years ago. This not a comfortable place for our mindset which is encouraged to be in control of our environment. The only solution, as I see it, might be to expand our awareness to look at factors that do not generally fall on our radar.

Mention above was made to the idea of benefit for the community. If we can make an attempt at reading the mind of those who lived when the towers came into existence, there is a strong chance that he or she, like us all today, was primarily concerned about physical survival and about satisfying the instinctual needs of all mammals.

There is an ever-present factor which should not be overlooked, especially in the Irish environment where such care was paid to traditional mores and custom. If the land surrounding a tower is reputed to be more productive and nourishing agriculturally speaking, as we learned from the incident of Callahan being ferried over to Devenish, that merits being investigated because it is possibly the main reason for their creation. When viewed from the purely practical standpoint and one asks, "What does a tower do?" There is no answer, at least not apparently, so we have no option but to reformulate the question from the context of the time they were built – well more or less, as we don't even know when that was.

Ireland is an immensely rich country agriculturally speaking, as was even apparent in the days of Diodorus Siculus (see the quote at the beginning of the chapter). Were the weather patterns so different then that two harvests of grass were possible, or was there something else going on?

There was no doubt a time when the country was covered by more forest than now, reducing the extent of open land to grow crops, perhaps putting a certain pressure for a need to produce on available space. This is all very hypothetical but the location of the round towers in open spaces, propitious to the growing of crops would indicate such an argument has some validity, especially when you consider the direction of the doorway, which is where the maximum magnetic force is directed, laying down a magnetic field which carries and can be measured from several hundred metres distance.

There is an additional element to the puzzle here. What if the towers connected to each other not only via the underground waterways, but on the surface? The density of their presence – and that is especially so in the central areas of the country on land conducive to growing. The energetic output from a tower as provided in Appendix II gives a reading for some of the standing towers. That reading is very probably much weaker now than when it was in operation, for the very simple reason the towers have all been revamped over time, or destroyed, or damaged, or no longer have their caps – either original or according to design, whatever that might have been.

That charge – magnetic or other, as can be demonstrated by experimentation, is the result of the combination of a structure over an energy source. For example if you take a conical form with a hole in the apex and then place the cone over an energy source, what exits the hole is a concentrated form of that energy. This is what is happening with the round towers built in the manner explained above, positioned over the crossing of underground water, then directed over the selected angle of country thanks to the "doorway" and the windows. This disposition has a double purpose, firstly to enhance plant growth, secondly to provide well-being thanks to the removal of the harmful influence of the crossing of underground water, and probably the dissemination over the countryside of a nourishing form of energy that might not have been there previously, at least not to that extent. In any case the danger so threatening in its effect against which the Chinese with their *feng shui* and, the Indians with their *vastu* warn.

CHAPTER 9
THE PURPOSE OF THE TOWERS

It is a commonly held belief that stasis in any environment, wherever it is to be found, causes issues that if not removed, results in problems. Whether it is in a human organism, or in a natural setting. The Chinese, for instance, in their traditional medicine go to great efforts to restore the movement of nerve, blood and essential fluids so to avoid blockages in passages that must be kept open if good health is to be enjoyed.

Observation of Nature has resulted over time in humans accomplishing the most remarkable feats in order to improve not only living conditions for ourselves and our domestic animals, but also enabling a healthier environmental state, as in removing swampland with the mosquito population, reclaiming desert by planting trees to retain water in the soil, and such like.

In a country like Ireland where water is to be found everywhere, both above and below ground, any method used to enhance its management could well be welcome, especially when it causes damage. Of course, the next question is "What damage?", and how do we even know that this damage is caused by the flow or the crossing of underground water. There is no ready answer, however, but given the feedback we have accumulated over time from hard experience and the knowledge base – for the believers – as seen in the chapter on geopathic stress, it seems quite clear that we can choose to believe that option, or not. Modern scepticism is a great boost for science and probably the force behind many discoveries, but at some stage one has to take a position and believe. I personally found that belief when losing my sight due to the influence of an underground stream, and in the subsequent search for a solution, that influence was removed. There is no way I can convince everyone, and absolutely no desire to do so, but my respect for the wisdom of our ancestors, and our lack of understanding in that regard, encourage me to at least give them the benefit of the doubt, if benefit is to be found. And what better way can I repay such a debt, in this modest manner, than by rendering unto Caesar.

If there is no branch of science that deals with such phenomena and the multitude of accompanying side issues of telluric forces, there is little surprise that any attempt to discuss such will be met with contempt and the accusation of pseudo-science. If researching into ancient lore, current problems and branches of neglected knowledge is pseudo-science, I AM a pseudo-scientist, and proud to be so. That might just be one way of bringing some relief to suffering. Unfortunately, access to the majority of "scientific" studies into geopathic stress have to be paid for, Elsevier and Jstor seem to have a fair volume of studies, that could well help us in any effort to learn, but experimentation and research in *feng shui* and *vastu* produce practical solutions rather than theoretical propositions, which are more in keeping with what I believe to have been the spirit and societal concerns of our ancestors.

Obviously, I am not suggesting the round towers are the solution for the mishaps water causes, but there is a strong chance that when all the towers were in their original state, complete with their caps and operating at full magnetic and energetic capacity, they were fulfilling a task the effects of which are still felt to this day in the immensely rich agricultural land that the island offers. What is more, the feelgood ambience found in Ireland is perhaps due to a certain degree to these intriguing structures. But the mystical aspect must be left for another day!

It is pretentious no doubt even to put forward a theory as to what the towers are for, but as you have read above, there have not been any convincing arguments put forward in the past, and given what we can imagine must have amounted to a complex and highly laborious task, we can reasonably assume that not only there was a motive, but that it was worth the effort.

There can be little doubt that some of the most convincing writing on the purpose of the round towers comes from the pen of Philip Callahan, a polymath, entomologist, biophysicist, infrared and low energy expert, traveller and immensely curious of everything as an avid observer of Nature. For those of you not familiar, he spent two years during the second world war manning a radar station in Fermanagh county, not far from Lough Erne, home to the Devenish Island round tower.

In several of his books, a) *Ancient Mysteries, Modern Visions*, b) *Exploring the Spectrum* and c) *Paramagnetism*, he explains one of the most practical aspects of the round towers, namely the mechanism of their probable operation and their operational method as wave receivers and transmitters, and he suggests some form of a boost to agriculture.

I would add a detail which he does not mention for the simple reason he was not aware of the factors that I believe are key because they are common to all authentic towers. The *HOW*.

How to achieve such a boost? Is it possible to achieve the necessary drawing capacity, as in the magnetic quality, to pull up that cocktail of frequencies developed by the two or more currents, and other, below ground?

The answer seems to lie in the fact that every stone in each and every round tower has a specific position, as a result of a deliberate gesture by the mason. Systematically they are laid with the negative magnetic polarity facing upwards, always on the outside walls of those towers that have not been tampered with, and most probably on the inside but it is generally impossible to access inside a tower to check that.

If Callahan had suggested the towers were operating as an energy-resonating system derived from the water below ground, rather than the aether above, then we would really have a working hypothesis – he didn't, but I do.

It seems timely now to try and put a little coherence into the story and explain what I believe what they were for and how they work.

The flow of water, like any movement or current, generates a magnetic field. Over time, that field expands affecting the surroundings with the frequencies produced. Due to the water capillaries or fissures in the bedrock or soil, or whatever the geological medium found between those currents and the surface, those frequencies if unable to find an exit, eventually rise to the surface of the earth. In probability that sets up a pathway to be followed by other frequencies as they develop, resulting in a substantial pattern of release on their upwards path. Those exit points were pinpointed by the ancients. Those frequencies are then released into the environment where they apparently do no good to the surrounding flora and fauna, to be slowly absorbed into the atmosphere. At the exit point, because there is a lack of resonance or some form of disharmony with the frequencies coming from below, the force is at its most intense, hence the damage that can be worked unless measures are taken to remove it. It would be a reasonable supposition that there is a very delicate balance in Nature that has taken millions of years to establish, and any upset in that balance can cause problems for some, and although perhaps there might be a benefit for others, we are at a total loss of appreciation one way or the other.

As explained above in the section on geopathic stress, one of the most harmful combinations of energies is created when two or more currents of water cross below ground, for the simple reason that they create a cocktail of frequencies which we normally don't deal with, and are not equipped to do so. When they reach the surface and work their particular effects, they upset the normal equilibrium, and dis-ease generally follows.

However, when that frequency cocktail is elevated beyond our immediate vital space, it is as if we are left in a vacuum, free of any interfering frequencies, where there is room for the individual energetic composition to express itself not only with less restriction, but in harmony with the surroundings. That neutrality is very much what one is aiming for when removing the frequencies from the area of the exit point. Again, in all probability conditions in our environment – temperature, daylight, pressure, the state of ionisation, and other factors all play their part in the overall result. Consequently, if one can operate on one or more of those factors, there is a chance that a comfort zone can be achieved.

Like the stone pillars of the cathedrals, the barrels of the towers concentrate (due to the layering effect of the oriented stones) and focus that energy from below carrying it upwards, releasing on the way through the doorway and the windows. Consider how much more efficient a hollow barrel would be, compared with a solid pillar, especially if you arranged a polarized array of stonework – as in the Round Towers, drawing that magnetic force field up, so that it works its modified and beneficial effect on the environment as it bursts through the door and windows – not only to the earth but the whole area, giving it a quite unique quality.

The arrangement of the doorways and windows in the towers would appear to create a spiralling movement, the force of which is released through the openings, and in the process somehow seems to remove the harmful effect of the combined frequencies from below as the 'pressure' is released. Is there a lumination effect operated, transforming that magnetic field into a negative ion complex? This is where the unknown component(s) lie(s) and the magic is worked. Whatever that effect is, it is going to have an impact on the surrounding area, and especially on the more sensitive components, such as the mycorrhiza.

One can easily imagine the effect of concentrating a source of energy such as the magnetic field of force coming from the crossing of underground water, by allowing that field to rise through the barrel of a tower which narrows, thereby further focussing the force, and narrows again four or five times on its way up to the top. By releasing the accumulated energy through the doorway and windows, the purpose of the energy flow is fulfilled and maintained at a constant rate. Such an arrangement would militate for a hole in the roof of the tower, or some form of release mechanism – perhaps the windows in the top storey accomplished that, although a hole in the cap would be a more consistent release, to ensure a constant and more regular flow of that force, as opposed to a sealed cap. It would not be difficult to perform some simple scale experiments using an energy-generating source, as I have done with a geometric spiral as the source and a series of five copper cones of 8 cm diameter spaced regularly.

The fact remains that whatever we call "energy" is in and around us all and it is modified by every event, however small or insignificant, but to what extent or depth shall always remain an unknown until such time that it either hits you in the face, or you knowingly devise some form of method or other to solve the imbalance. If we unconsciously control our existence by breathing, digestion and circulation how much better a job we could provide if we were to assume control of the few factors that we can control, perhaps what the ancients were doing! You know the benefits you can achieve by breathing or relaxing muscular tension because of the impact on the emotions and thought, so why not apply a round tower?

Wilhelm Reich (1897-1957) in the 1930s was able to objectively measure the movements of a bio-electric charge which he named orgone by using a very sensitive millivolt meter with sensors attached to the body to record subtle bio-electric charge. He found the energy flowed from the inside body core to the outside surface (towards the world) when a person felt pleasure or expansion; and conversely, it flowed from the surface to the interior (away from the world) during states of anxiety, fear, and contraction. There seems to be room for somewhat of an analogy between the energies coming from the earth when released into the atmosphere.

Reich also noted that the conditions of expansion and contraction affected a person, not only emotionally, but down to the autonomic nervous system, to the cellular, and even chemical levels. States of expansion produce parasympathetic conditions associated with dilation of the blood vessels and increased circulation, pain relief, better digestion and peristalsis, lower blood pressure; and the stimulation of potassium and lecithin production; along with creating a sense of well-being, and sexual excitement. States of contraction, however, produce sympathetic effects: constricted blood vessels, less blood flow, and often pain. In addition, the contracted condition increases blood pressure and heart beat rate, adrenaline flow and cholesterol; it inhibits digestion and blood supply and is associated with the emotions of anxiety and "stress". The ability of the body to expand and contract and not become "stuck" in one mode, created what Reich called the pulsation of life which distinguished the living from the non-living. This pulsation of expansion and contraction also followed a specific rhythm.

If we are ever to resolve this question once and for all, especially to the satisfaction of the scientific-minded, I think it would be a matter of manufacturing some simple instrumentation that is capable of measuring the magnetic force field at 1) the exit point of an underground water crossing, 2) in the centre of the Tower, 3) at the doorway, 4) at the windows, and then at varying distances from the Tower. I take those measurements using a pendulum and a chart of my own making.

So long as the tower is located over the crossing of underground water, it does not matter what type of stone is used, but each and every stone must be laid with the negative magnetic polarity facing up, so that they create the necessary magnetic thrust to draw those frequencies up and away. In all probability there is a connection among underground waterways, thus enabling a form of pressure-release allowing the water to stay closer to the surface, giving the Emerald Isle its unique properties.

Consequently, it does seem logical to question how the Roman Catholic church can claim any involvement. It would not be very subtle to make a claim unless one can provide some kind of proof, or even better, a reason for doing something. That would be dishonest and not go down well with the people who did actually make the towers. But if those people were no longer around, what can stop the church claiming responsibility for the work, not in so many words perhaps, but by systematically placing graveyards and other ecclesiastical buildings in the immediate vicinity, as well as encouraging those with influence to dupe the public. Unfortunately by doing that, the whole purpose of generating a magnetic field of force conveying a benign energy over a wide area is defeated, and that benefit is squandered.

Having laid the case for your consideration, it is now time to go into greater detail about the various principles which the ancients seem to have mastered and applied.

PART III
THE THREE PRINCIPLES

Facts are more powerful than words.

This section of the book will take a closer look at the evidence we still have at our disposition, and hopefully in the light of what has been explained above, make it easier to consider – perhaps even understand – the methodology employed. In a manner of speaking, we have to work backwards and reverse engineer the how and why because we have no user manual for this phenomenon we seem to be dealing with. I will start with the youngest in the series.

If you have ever visited any of the cathedrals built in western Europe during the 12th or 13th century, you will have perhaps experienced a unique sensation of calm that is not found in any other ecclesiastic structure. A sensation on walking into the building that a weight has been removed from your shoulders.

That, I believe, is quite normal for the simple reason the techniques employed were unique and never to be repeated again since those times.

As explained in my book, *The Way of the Skeptic*, vibration is a phenomenon caused by the motion of periodic or random frequencies generated by an energy source, for instance insect antennae, emotions, the heart, one's voice, electric (infrared) electromagnetic impulses, sounds, and so on. A less obvious, but equally potent, source of vibration is architecture, with its volumes, proportions, and consequent projections. (This is where Callahan could well have been close to the truth, but rather than outward-facing antennae, the towers are inward-facing – to what is below the ground).

Whatever form the vibration takes, naturally enough the movement interacts with the surrounding environment, and especially the human complex as it is composed of zillions of cells all vibrating at as many frequencies, or so we are told. One can assume that the resulting chaos is uncomfortable and any relief provided by whatever means might be welcome, but I digress. According to modern physics, magnetic polarities are created as a result of the flow of this energetic movement, thereby producing atomic structure with its complex of particles.

What if the heart's beat was the key to human, indeed mammalian, life? The heartbeat might be an electric impulse, with its origin in the combination of radiation from the cosmos, the earth, and the environment. Or perhaps we are ambulant thermoelectrets, producing a constant vibration: a resonate cavity pulsating thanks to its expansion and contraction, perhaps the conduit for a host of information coming from the outside via contact with the skin and the senses, both receiving and transmitting. It does not seem an unreasonable proposition; it clearly does more than keep our life blood circulating. We know that an antenna does not have to be made of metal—both insulative (dielectric) and conductive materials can fill that role, and in fact an antenna made of dielectric materials performs even better than a metal one in the one-centimeter range.

No matter the explanation, our actual experience provides the best, and only, guide. When we awake from sleep, we move out of that state of blissful restoration – so long as we do not sleep over geopathic stress – found in deep sleep or in conscious immersion into what is known as the parasympathetic nervous mode. The waking state is dependent on the sympathetic nervous mode—the flight or fight response, which brings a number of reactions with it such as activating the brain, the muscles, the insulin pancreas, the thyroid, and the adrenals while at the same time stimulating certain hormones and substances, insulin, cortisol, dopamine, serotonin, the thyroid hormones, and a host of others. While we are very good at differentiating and defining these substances, we do not know much about the states, reactions, and effects generated by their interaction with other substances and with our environment. This is a great loss, because we send out vibrations reflecting our internal condition, which could potentially have an impact on everything around us; and in turn there is a knock-on effect on the environment, subsequently feeding back to ourselves. We could learn much from these interactions, if only we were truly in tune with ourselves – and our environment, and this is what I think the ancients were driving at with their science.

While on the subject of vibration, there seems to be a commonly experienced isolation felt in certain places throughout the world. This is an inevitable consequence of the buildup of all sorts of external vibrations in the form of electromagnetic radiation, especially noticeable in shopping malls and motor vehicles. Winfried Otto Schumann was the first to study the theoretical aspects of resonance from the earth's ionosphere, although an Irishman, George Francis FitzGerald, had put forward the idea in 1893 that the top layers of the atmosphere are good conductors of electromagnetic waves. It is a recognized fact that anyone travelling in space suffers from jet lag and disorientation, the reason being that in the ionosphere and beyond there are no longer the same waves we enjoy here below on the earth. When deprived of these frequencies because no longer connected with the earth, the human metabolism is upset. In the same way, when you visit the shopping mall you are surrounded by an intense curtain of electromagnetic radiation coming from every which way and from multiple sources, which deprives the atmosphere of negative ions and supercharges it with positive ions.

We live within an electric field (well that's how we choose to term it) that varies according to climate, location, and individual height, with a negative charge at ground level potentially rising by 100 volts per meter (up to higher altitudes, when it drops off again). So, at six feet tall (one meter, eighty centimeters) you have at your nose's disposal 180 volts; this is the earth's naturally occurring field. In a car—which is a metal box—you are disconnected from the grounding capacity of the earth, more or less, because of the rubber tires, and as a result are subject to road rage, greater fatigue, and other effects because no longer in contact with the potential provided by the earth's frequencies from which one is disconnected.

Now, what if one combines this understanding with a technique that can harmonise vibration?

The re-discovery of that technique, or methods perhaps, has been written and speculated about, albeit obliquely, in Louis Charpentier's book, *Les Mystères de la Cathédrale de Chartres* (The Mysteries of the Chartres Cathedral), and what a fascinating intrigue it all makes.

Basically, the story runs as follows:

Saint Bernard de Clairvaux (1090-1153), a Cistercian (the monastic order renowned for their affinity to water) apparently learned of a secret concealed in the Temple of Solomon in Jerusalem (the British were still digging in the Temple during their occupation between the world wars). Quite what the secret was, history does not tell, but Charpentier tends towards the Ark of the Covenant. Whatever, Bernard rapidly set up (in association with Hugues de Payens) a monastic-soldier order with a group of very influential nobles from the Champagne region to bring the secret back to France. It must be remembered that all the knight members of this order were of 'noble' birth (families of influence due to their financial and political clout). Fascinating though the plot may be, it is not our concern and the secret disappeared even deeper with the disbanding of the Knights Templar (the monastic-soldier order in question) in 1309.

There was, however, the most remarkable consequence to the order of the Knights Templar being established, and the impact of the 'secret' they brought back to France (or what we know as France today).

For a period of almost two hundred years, from about 1130 to 1310, a Europe-wide cathedral-building campaign was maintained, with construction in France, Germany, Portugal, Spain and the UK. The most rapid and widespread construction programme of cathedrals ever seen in the history of man, what is more, using a totally unique style of architecture, the so-called Gothic, with innovative techniques never employed to-date, and never since.

The extent, intensity and precision of building has never been repeated. By way of example, in France alone between 1150 and 1250, one hundred and fifty churches were commenced.

It is no easy task to grasp what a huge undertaking the building of a cathedral is. Even when you are in the centre of the labyrinth inside Notre-Dame de Chartres, looking up to the roof, the immensity and overwhelming sensation tend to make you feel very small, and a rather special mindset and/or imagination has to be forthcoming. Twenty-five to forty odd years is needed to build such an intricate structure. Even with a company of specialists and an army of labourers that is not a long time. A logistics nightmare with people who need to be paid, housed, fed and organized with clockwork precision, not only because their services are costly but they are required on other cathedral worksites, like fifteen in the 12th century alone in Normandy.

One must reasonably assume that the Templars had access to plenty of cash, thanks to their international network perhaps. The Roman church was rich even then, but not that rich and very probably could not have financed all these construction sites. Once again, rumours abound – silver from South America traded for wool, a new economic model,.... Let alone the political considerations, for there seems to be a considerable probability that there was a major conflict between the two parties, which the Templars lost. One of the consequences of that struggle concerns us directly here, inasmuch as the loss of this architectural principle disappeared with the Templars.

It is now time to return to our original quest, for this incident is but a stone in the path, albeit a major stepping-stone bathed in mystery, setting the tone to the death struggle that we do not hear about in polite society when mention is made of the church.

Whatever the truth, the cathedrals were erected across Europe, even in the UK during this period – ALL using the same principles. For convenience's sake I shall refer to these as:

a) The Cross Principle, no nothing to do with that appalling method of dealing death, but the crossing of underground water, and;
b) the Geographic Principle, whereby one obtains the measurement to be used as a proportion in the construction of the building in question. There is a third principle found in all the older buildings worldwide, but NOT in the cathedrals, and this seems to have been replaced by the Geographic Principle to accomplish the same results; that is
c) the Polarity Principle, which we shall encounter when dealing with the round towers.

Architecturally speaking, while the Gothic and Roman styles appeared to be concomitant and continued over the next few centuries, never again was this Geographic Principle, found uniquely in the structures built during the Templar period, to be used in any form of architecture after the disappearance of the Templars.

That is only partially true, they were the secrets held by the Templars, but not theirs by any means, otherwise they might not have come to such a sticky end. If the church were at the origin of what I refer to as the Geographic and Polarity Principles used in gothic architecture, they would have used them in subsequent construction, but that was never the case, and the pope sided with Philippe le Bel, the French king who gave the order to do away with the Templars, due to political expediency rather than any intellectual conviction in removing them.

If the Templars had perhaps revealed that knowledge and its provenance, rather than keeping it under wraps, it would not have been the subject of such jealousy. No, the secrets 'found' by the Templars dated from a previous age of which we shall probably never learn more.

As can be seen in Rosslyn Chapel, a great deal of symbolism is revealed by the early masons who built that remarkable structure, but not the Geographic Principle, so it would appear that the freemasons were not privy to the secret, even at a period so close to the disappearance of the Templars.

Incidentally, the last stand of the Knights Templar as a fighting unit was probably at the Battle of Bannockburn in 1314, which saw the English front line shattered by a cavalry charge. That should be cause for surprise because the Scots at the time of Robert the Bruce did not use horses in battle, unlike the Templars.

The Templar order was disbanded in 1309, after the last Master, Jacques de Molay, was burnt at the stake in 1307. Construction of the great cathedrals ceased in Europe around that time.

Jacques de Molay prior to his horrendous ordeal, while not at the hands, but on the orders of Philippe le Bel, had spread the word that the soldier-monks would be received with open arms in Scotland – Portugal of course became home to many of them, but Freddy Silva in his excellent book, *First Templar Nation*, explains that all very well. The fortunate few who did escape did not apparently include any of the brethren privy to the building secrets. The reason for saying that is Rosslyn Chapel, astonishing though it is, and perhaps one of the finest works of art of a newfound masonic society, does not incorporate either the Geographic Principle, nor the Polarity Principle although the building is located over the crossing of underground water, but that is common to all ancient sites which were usurped by the new arrivals, i.e. the church.

CHAPTER 10
THE GEOGRAPHIC PRINCIPLE

Despite the major modifications made in recent times to many European cathedrals and monuments built during the Middle Ages, the effects of what I call the Geographic Principle are demonstrably active even today.

There are two main elements to the Geographic Principle:

Orientation was always a major principle in the siting of a building in the past. An obvious factor being the rising sun and our dependence on that body for our material existence, but it is far from the sole criterion. Chartres cathedral, amongst the churches is a notable exception, with its orientation towards the north-east, perhaps in alignment of the direction from which one of the main underground water streams flow.

However, before one orients a building, and before finding the right place where it is to be set up, one needs to be quite decided on the fundamental purpose. The story put out over the centuries that the church was at the origin merely stops us from considering further – WHAT WAS THE PURPOSE?

Can we seriously accept that forty or so years of construction work, at vast expense of labour, money and expertise, was to build a new style of place of worship?

That is what we are TOLD, and we believe it without any further thought.

What if the crossing point of the underground streams over which every cathedral was built in that period can be turned to advantage, one could then make as if the church were at the origin of the resultant well-being?

No matter if the pagan sites which apparently existed beforehand in those same localities used their own methods to remove the harmful effects of the water crossing, the memory of the general public has never been too solid, especially when you provide something bigger and better.

That makes for an extremely expensive marketing programme, but what better way to demonstrate power and seize authority over the common mortal – after all, your future investment. If, alternatively, the Templars paid for it all and were then conveniently 'disappeared'..... But that is not the subject here, but definitely an insinuation!

I would maintain here that the emphasis on the objective – and subsequently usurped by the Roman church in this instance – was for the well-being of the general public, because once one has achieved such a sensation of comfort, even in part, one potentially gains considerable control. If you can offset the harmful effect of the most powerful telluric influence by conveying its force upwards to the heavens, one can make the surrounding area on the surface of the earth welcoming and calm. That must have been, and still is, a promising accomplishment and likely consideration to be borne in mind as a reason for that objective.

But that is not the end to the matter, one also needs to know the method to achieve that noble aim of well-being. In the case of the cathedrals as we can deduce today, that required a precise knowledge in order to calculate the 'cubit' of the locus. That is my livelihood so you will excuse me for not sharing it freely here.

The science that the Templars brought back with them was vast with potential, so there was probably some interesting reading stashed away in Solomon's Temple too. I would even venture that never since the construction of Kheops pyramid has so much wisdom been potentially revealed. Are you aware that the British during the period between the world wars when they were "protecting" the Holy Land, were digging extensively in Jerusalem?

The Chartres 'number' is 73.8 centimetres, or according to Charpentier 100,000th of the total distance longitude around the world on the same latitude where the cathedral is positioned. I cannot agree with this calculation, but playing with it allowed a more precise method to be revealed. This serves as the basic measurement, or the cubit of Chartres, that is combined, applied to and proportioned throughout the construction of the building. This same geographic principle is applied and used for each and every cathedral built during that period when the expert craftsmen or companions were practising their skills.

While there is no record of how these craft-guilds came into being, it was most probable that was the form of schooling that developed. It is hard to imagine how else one could produce such a large number of skilled workers in so short a period of time covering such new and varied trades as stone-masonry, carpentry, roofing, foundation-laying, scaffolding, glass-work and colouring, rose-window stone lacework, flying buttress techniques, and so on.

The Geographic Principle, as I call it, is one of the keys to the mathematical computation of the structure, as well, I firmly believe, as the cause of the sensation of well-being developed when using that measurement. The reason for saying that is because this sensation of well-being can be easily reproduced when using that measurement specific to the geographic location of any position in question. This very same principle appears to have been used in just about every ancient building throughout the world, whether in Egypt, China, Peru, Mexico, Greece or France.

That would imply that the knowledge required to implement this technique is old, much older than the Templar institution, older than Greece, and probably antedating Pharaonic Egypt. More questions which must remain in suspension, and even if we did find archaeological evidence at the bottom of the seas, we would still be none the wiser. So, let's stay in the domain of the possible, practical and what we can physically work on.

It is of some consequence to learn that round river rocks, boulders, were discovered in the foundations of Chartres cathedral during renovation work in the 1960s; regrettably no mention of magnetic polarity was recorded. Of course many of these European cathedrals and churches were built on older pagan sites, where remedial work had already been performed, consequently benefitting the site thanks to the measures taken, in addition to not having to prospect for new sites.

CHAPTER 11
THE POLARITY PRINCIPLE

It is now time to introduce the second principle, which adds one more, and perhaps the most significant, ingredient to the brew, for it is here with the round towers that we find combined both the Cross and Polarity Principles. Perhaps even the Geographic Principle if one takes into account the dimensions of the towers. These could well include overall height, positioning and height of the door, windows and openings, height above sea-level, orientation, and so on.

If a stone has a magnetic field of force, it is only a simple step to experiment with it as a function of its environment, make-up and position, to achieve the benefits that are so manifest when applied.

We are familiar with the benefits to plant, animal or human by modifying their current state of well-being or health when performing the laying on of hands, magnetism, Reiki, or chiromagnetism as I prefer to call it – all examples of what man has always known and used throughout the ages. Could it not be reasonably assumed that this applied form of magnetism can be achieved by other sources of magnetism, not just via the human medium?

Is this not the same force emanating from standing stones? And of a much greater potency. By its very nature, it would be a constant, with less variation than the field coming from an individual's hands. It would also be, in all probability, of a greater or lesser intensity as a function of its positioning in relation to the telluric influences over which it is placed, therefore providing a permanent output which will be felt by anyone or thing in its proximity.

That is perhaps the principle our ancestors understood and worked on when they erected the stones, whether obelisk, menhir or round tower.

It seems that certain stones have a 'healing' capacity, maybe due to the magnetic field providing a harmonious energy. As recounted above such stones can be usefully employed to restore health – in certain circumstances, but what is more intriguing is the re-establishment of energetic balance in the more general environment, which seems to be of equal importance in maintaining well-being, which then contributes to a more harmonious physical, mental and emotional state. In much the same way that magnets when applied to the skin over a broken bone, the healing of the bone is faster than without them, generally the inflammation is more rapidly absorbed, with reduced pain, and a happier patient.

Communal well-being is clearly not a concern in modern society, but we might well be heading towards a time when that could be a useful addition to environmental policy. In the past, even in living memory, people would regularly gather together and profit from the communal buzz created around an event or location. The sense or meaning of the event was often lost in distant history, perhaps appearing in mythology, but generally surrounded by such vague perceptions of the original purpose, that it loses any significant sense, but it is still celebrated nevertheless. It would be so much more logical if measures were applied knowingly to achieve that well-being for one and all.

That is the aspect we are now going to look at a little closer.

Would it not be too much of a coincidence that the negative magnetic polarity is systematically positioned towards the sky? If it was not a coincidence, that would indicate that such positioning was a purposeful architectural application of a recognized phenomenon. As previously mentioned, this same disposition of stones in building, negative (-) above and positive (+) below, is found in numerous structures around the world, as shown in Appendix 3.

So, the idea is that they were aware of the positive and negative magnetic polarities in stone, because they knew that those properties, specifically in paramagnetic stones, were powerful stimulants – for agriculture, animals and humans.

We could usefully look at another monumental vestige on our journey, and take one step further back in time.

THE OBELISKS

Ancient monolithic obelisks distributed around the world all have something in common with the round towers, they share the same magnetic disposition. In other words the negative polarity is at the top and positive polarity at the base, however and needless to say, no one has noticed or commented on that fact. We are led to believe that there are 28 obelisks in all which came from Egypt, of which only 7 are still in their country of origin.

It is said that the obelisks were generally erected in pairs, as one finds in Karnak, but that is a later development and possibly not the original disposition. Why would one have two obelisks side by side, when the protagonists were magnetically inimical? That may be the case if they were too close, but if the obelisks were spaced in full understanding and knowledge of the forces one is dealing with, there is a strong likelihood that a magnetic field of force of a specific kind can be achieved. That would seem to be confirmed with the dolmens or standing stones. Some of which appear to have been used for healing purposes, as tradition would hold in Ireland, and perhaps elsewhere.

Incidentally, the tallest obelisk in Egypt is to be found in Karnak, originally erected by Queen Hatshepsut, who was much maligned by her successor who removed many of the monuments erected during her long reign.

Most of the obelisks were inscribed, sometimes on all four sides, and in one or more (four columns on the four sides for Seti II in Karnak) columns of hieroglyphs on the side. There are, however, a few ancient obelisks with no inscriptions, including the 'Vatican' obelisk which Caligula brought to Rome. Rome today has more Egyptian obelisks (13) than those currently above ground in Egypt! What did they know?

It is not practical to engrave when you are perched 60 feet up in the air, so it would be reasonable to assume that they were inscribed whilst still on the ground. The information of the inscriptions is very relative to the reign, the title and praise of the person concerned, although there have been additional columns of hieroglyphs added in later times – upto four hundred years later. But if they are not all inscribed to commemorate or extol the feats and works of the person commissioning the monument, why would you go to the pains of quarrying such a large piece of masonry?

As is so often the case in Egypt, as elsewhere, there is more to what one sees than meets the eye. The obelisks are a perfect example, once we move beyond the symbolism, the grandeur and egocentricity. There is perhaps a grassroots pragmatic function, maybe to evacuate the damaging effects of the crossing of underground water, thereby 1/ creating a space of well-being for whoever spent time in the neighbourhood? Or 2/ with an antenna releasing and, in a manner of speaking, broadcasting vibration from below ground, or 3/ drawing in energy from the atmosphere at a certain height?

1/ is achieved by the attracting force of the stone obelisk to pull the harmful-to-human magnetic field caused by the combined frequencies of the water crossing upwards and away into the atmosphere. This same principle is applied when building a pillar in a cathedral or church.

2/ and 3/ would be achieved by a concentration of negative ions surrounding the negative polarity of the obelisk.

An obelisk as well as a round tower can achieve all of these functions.

As is to be expected, there is controversy as to the origin of the Egyptian word which might define an obelisk. We have the term TEJEN, which in Egyptian signifies protection, or defence. TEKHENU, meaning to pierce. This last is reminiscent of the Thai *chofah*, that part of a Buddhist wat (temple) or palace roof architecture where a decorative Garuda-type addition 'pierces the sky' (Sanskrit *chho*, to cut, pierce; Thai *fah*, sky).

Why protection? Against what?

If my proposition is acceptable, the answer seems obvious – against the harmful effects of the crossing of underground water. There does not seem to be any other logical explanation, but as always, I'm all ears.

The problem, and it is one of consequence, is that there is only one obelisk in Egypt still in its original place, or so we are led to believe. The answers using a pendulum as to whether that is actually the case, and if the obelisk is over the crossing of underground streams, are both affirmative.

Obelisk of Sesostris I, 12th Dynasty, Middle Kingdom in Heliopolis

The obelisk of Sesotris I (aka Senusret I, Sesonkhosis or
Senwosret I)
(Thanks to Google Earth)

There is a crossing of subterranean currents below this obelisk, even after three thousand-odd years since its erection. It is the oldest known obelisk according to the paraoh.se website.

The comments below taken from the internet are listed to demonstrate our amazing imaginative ability, as well as a more concerning issue, our serious lack of gnosis:

"Originally created as ancient Egyptian shafts of 'frozen' sunlight..."

"Obelisks—giant standing stones, invented in Ancient Egypt as sacred objects—serve no practical purpose."

"A bizarre Egyptian luxury object... erected to commemorate an individual or event and honor the gods..."

And at last, some sense: "Obelisks, everyone seems to sense, connote some very special sort of power." Regrettably, the sense stops there!

Why not and who is to say? But if, as is the case, other stones erected around the world reveal the same negative magnetic polarity up, and positive down, why is no attention given to this FACT, and the emphasis placed on non-sense?

If we are attentive and know where to look there is evidence throughout the world that former civilisations were not only aware of these issues, but took active measures to ensure that these telluric influences were attenuated when overly aggressive – or avoided. What is even more pertinent, profited from when it was possible to enhance the benefits. For it is possible to transform those harmful effects and turn them to benefit.

In Ireland alone there are between 45 and 50 thousand stone circles according to the excellent Jack Roberts, *The Sun Circles*, 2013. It is not just the stones in the circles that demonstrate this magnetic polarity, the towers do as well. The difference being that each and every stone in a tower is positioned in such a way that the Polarity Principle is achieved, and thereby multiplying the field of force in consequence.

One step yet farther back, perhaps beyond.

CHAPTER 12
THE CROSS PRINCIPLE

Nearly 97% of the earth's usable freshwater is underground apparently, and 73% of the surface of the earth is covered by water. That leaves a lot of room for investigation and research, but does give us an indication of the importance water plays in our biotope.

As explained in some detail in the chapter on geopathic stress, one of the most harmful conditions found for human and animal life is the crossing of underground water. The damage caused by this phenomenon was known and considered so detrimental that in the Chinese tradition, one does not build a house over such a geological event. That presupposes that it was a known danger to be avoided because problems had arisen at some stage in the past, and furthermore, someone knew what the origin of the trouble was, and most importantly, how to locate it.

No ancient site is ever found built on such subterranean aquifers. Medieval builders would reputedly even test their sites first by putting sheep there and observing if they avoided the area before building, and it is well-known in agricultural communities that animals will not give birth over such spots. It is only in the last forty or so years that attention is paid to this problem, and only in Germany (among the western European nations at least), is it necessary to acquire clearance from a recognised dowser before building permission is granted. Naturally, that clearance includes more than a site being free of the crossing of underground water.

Underground streams are made up of water which flows through any underground passage, fracture or fissure. As it does so it produces its own electromagnetic field often high into microwave frequency, as well as generating flows of ions, generally positive ions. This field will fluctuate depending on what is dissolved in it, sewage, graveyard or dumping ground sediment, fertilizers and insecticides, how fast it is flowing or whether it is interacting with any other type of earth energy. Interference with the earth's natural energy field is particularly marked by the flow of these underground streams especially where two watercourses or other types of energy line cross. In a league table of radiation associated with biological damage, the outside line of a subterranean water course would be at the very top.

These are the natural but evolving entities that we are dealing with, albeit with little, if any, real understanding and regrettably, with a widespread lack of concern, as they stream into our environment that is already cluttered by vibrations emanating from power grids, electrical circuits, computers, mobile phones, faxes, amplified by metallic structures and furniture, all of which play havoc with our vibrational balance. There stands the individual human being struggling in their midst with very little help coming from the cosmic energy that could balance the whole, if we were in an ideal setting.

When you realize that the round towers, as well as many other old structures are built on these hazardous-for-human zones, one cannot but wonder what made the builders reckon they had some kind of remedy that could avoid the danger, and additionally, improve the situation. This would indicate that they not only knew of the problem but they had the expertise to use that energy form and transform it into something beneficial. That deserves the name of healing if anything does. Using a source of energy, drawing it up and then directing it to where it is no longer causing liability but turning it into an asset for a larger community.

CONCLUSION

If such a thing is possible?

The search for truth is obviously a very subjective matter, and in the final analysis, a very egocentric one, because we are all determined to be right, and to a certain extent we are. It is the extent, the scope which is judged by others and over time that determines the validity.

Obviously, the truth with regard to a subject can only be found when all evidence is collected and assessed in the context of that subject. If any aspect is uncertain, out-of-place or inconsistent with the evidence and cannot fit into the original context, it should be considered as such and afforded the merited judgement. If all aspects of a subject are considered and found to be in harmony, it can be considered as probably right, and as close to the truth as we can determine with our parameters limited and biased by time and context. Hopefully, thanks to a wholistic approach of thorough investigation of what is accessible we can determine what is right or otherwise, leaving to one side human opinion, and much more our tendency to deride when we do not agree with what others suggest.

A second point, and it is perhaps the most important in this essay because it concerns how we assess the evidence, having achieved and registered awareness and understanding of something. In this instance and especially pertinent to this essay, an awareness of something invisible, subtle and for which no instrumentation, even if it existed, can prove how it impacts the human ability to know, or gnosis. As a consequence, when confronted with a situation where it is not possible to know the causes, rather than falling back on opinion (generally of others), I can only ask that you look impartially at the empirical <u>effects</u> so as to link any possible causes, in view of what is evidentially available. Furthermore, I would ask you to consider the possibility that kinesiological methods (radiesthesia in my case) are valid in determining not only the existence of these hidden forces, but their effect, the latter thanks to an ability to employ our intellectual capacity, experience and intuition. I am not asking you to cast your cerebral logic and intelligence to the winds, but simply to consider the possibility of the existence of such methods in helping us discover the nature of these subtle forces in light of their effective impact on our subject here, and maybe on all that we know as life.

APPENDIX I

REFERENCE BOOKS WORTHY OF THE NAME

There are a number of books which would be worth reading to obtain a less biased view of the towers than the ones mentioned here. Starting with the rector of Ennis, Marcus Keane, who wrote a most useful book *The Towers and Temples of Ancient Ireland: Their Origin and History Discussed from a New Point of View* in 1867. For a man of the cloth, Keane is way out there, investigating and discussing with numerous literary references and pictures in support, a whole range of ideas in the aim of discovering, not imposing. How refreshing! He mentions on page 305 that lists of round towers amount to one hundred and twenty, with the remains of sixty-six or so remaining.

The Round Towers of Ireland by Henry O'Neill, Part I, 1877. A very brief, 56 pages, but useful study on the towers in the Dublin area – Clondalkin, Lusk and Swords. Unfortunately, the second part never saw the light of day.

The Round Towers of Ireland or The History of the Tuath-de-Danaans, by Henry O'Brien, 1898, 551 pages. This is the author mentioned in Chapter 6 who produced an extraordinarily rich document with a very varied reference base, proposing that the towers were fertility or phallic symbols (*lingam* in Sanskrit) employed in an ancient form of the veneration of fire or sun and moon worship introduced by Asian Indians.

The Mysterious Round Towers of Ireland, by Thomas Sheridan, 2018, 125 pages. A very eloquent plea to take a new approach of the towers and to move away from the prevailing assumption and theory upheld by academia and the church. His arguments are convincing and forceful, particularly the question as to why the towers were not destroyed by the early Christians as pagan symbols.

APPENDIX II

SOME PERTINENT DETAILS OF SELECT TOWERS

Grateful thanks to K.Schorr, the creator of the ex-website roundtowers.org from which some of the info is drawn.

All the towers listed below are over the crossing of underground water. So it would seem to be a pertinent question to ask what is the energetic score of the field of force at that locality. This is what I refer to as the Radiesthetic Reading (RR). The extent of that force depends on a variety of factors, which we may be able to assess or not, as a function of the time of day, other input and obstacles (power lines, phone mast, etc.), temperature and atmospheric pressure, and such like. It is provided for info and obviously using my dowsing capacity, chart of my manufacture, and limited understanding.

Using the Universal Pendulum with 6 hemispheres, as devised by the two French radiesthesists, de Belizal and Chaumery.

Geographic coordinate (GC) – Locality and altitude when available.
Elev – Elevation above sea level.
Ø – diameter of the tower at the base.
Tower material where available.

Number of openings and direction when available.
Height of door and direction when available.
Height of tower when available.

Co Antrim

Antrim Round Tower - Aontreibh "single ridge"

GC: 54 43 26 42 N 6 12 31 47 W. Elev 35 m. RR: 40 Ø 4.80.

Height: 28 meters above ground level. The NE facing doorway is approximately 2.35 meters above the top offset. Each is comprised of a large slab of granite, as are the four sidestones of the doorway, in contrast to the rough local basalt rubblework of the rest of the tower. The windows are all lintelled with the same rough stone as the rest of the tower and most are fitted with simple wood frames and glass. The top storey windows face the traditional compass points (almost) and are smaller than the other windows in the tower. The other windows, in ascending order face ENE, S, W, and again S. 8 windows in all. Surroundings cleared in 1800.

Other Items of Interest: A large boulder, with two sizable bullauns, lies approximately 6 meters from the tower, slightly to the left front of the doorway.

Armoy Round Tower

GC: 55 08 05 N 6 18 37 W Elev 108 m. RR: 45. Ø 4.6 Circumference:14.48

Laminated schistose sandstone (mountain freeze). The south-facing doorway is just 1.6 m above ground level. The very narrow doorway arch is cut from a single massive lintel with a decorative raised moulding. The tower is currently 10.8 meters tall. There are no surviving windows. Excavated in 1843.

Ram's Island (not considered as a round tower)

GC: 54 35 06 N 6 18 19 W Elev 22 m. RR: -2. Ø 4.16. Circumference: 12.1

Capless tower of rough masonry, 12.8 m high, external diameter 4.16, three storeys. Door facing SW 15 cm above ground filled in c 1940. Original doorway SE c 4.60m high.

Co Carlow

St Mullins
GC: 52 29 20 N 6 55 38 W Elev 17 m. RR: 20. Ø 5.1 Circumference: 16.08
Little more than a foundation, this roundtower stump is just 1 meter in height with no doors or windows. It is evenly coursed of granite blocks dressed to the curve both internally and externally with rubble between the walls.

Co Cavan

Drumlane
GC: 54 03 29 N 7 28 44 W Elev 57 m. RR: 45. Ø 5.09 Circumference: 16
Incomplete with level top, pale brown limestone. 11.6 high, for the first 6.7 the stones are well cut and closely fitted, the remainder are poorly course and rough. Door facing SE at 2.74. One window directly above the door, 7 m high. Excavated in 1844.

Co Clare

Drumcliff
GC: 52 52 05 N 8 59 50 W Elev 19 m. RR: 53. Ø 4.88 Circumference: 15.34
A ruined stump, the base of the tower is approximately 4.9 m in diameter and it stands 11 meters above the lowest ground point. On the lower and more level south side, the tower reaches a mere 2.4 meters. There is no door, and the windows no longer exist, although a report of 1808 mentions the door at 6 m high and three windows, one facing west, another east.

Features: While there are no doors or windows in this tower, the large breach offers a fine view into the construction of this tower. Most of the courses of stone are quite even and the interior stone is dressed quite smoothly. The traditional "sandwiching" of an outer and inner wall filled with rubble is quite evident, as are the floor corbels on at least one floor.

Dysert O'Dea

GC: 52 54 34 N 9 04 06 W Elev 24 m RR: 38. Ø 5.89 Circumference: 18.52

Well coursed large limestone blocks, dressed to the curve. At the base, the external circumference is 18.5 meters, one of the widest of the recorded towers. At its highest point, it rises to 14.6 meters. The wide doorway, facing east, 4.46 high is unembellished and arched with six blocks, running right through, the keystone slightly protruding. The left jamb is comprised of three stones, and the right has four stones.

Iniscealtra Round Tower - Holy Island

GC: 52 54 56 N 8 26 53 W Elev 40 m RR: 50. Ø 4.58 Circumference: 14.4

The tower rises 22.3 meters above the present ground level. The ENE-facing arched doorway is 3.0 meters above the ground. The three stones in its arch run the entire depth of the doorway. The north-facing angle-headed window at about the second floor level, 8 m high, is especially well cut with finely dressed stone. Three other lintelled, square-headed windows facing ENE, SSW and NW. Excavated 1976.

Killinaboy Round Tower

GC: 52 58 13 N 9 05 05 W Elev 40 m RR: 40. Ø 5.08 Circumference: 15.95

A featureless stump of roughly coursed limestone

Scattery Round Tower

GC: 52 36 51 N 9 31 00 W Elev 9 m RR: 47. Ø 5.08 Circumference: 15.95

Unique doorway at ground level, corbelled in three stones, ESE. 26 m high. 9 windows, N, S, SE, WNW, N and four slightly larger at the top facing NE, SE, SW and NW. The first window is immediately above the doorway. Repaired in 1855.

Co Cork

Cloyne

GC: 51 51 43 N 8 07 13 W Elev 25 m RR: 50. Ø 5.17 Circumference: 16.25

Purplish sandstone, dressed to the curve. Door 3.4 high, 9 windows. 30.5 m high. Excavated in 1841. Vaulted repaired and replaced with battlements 1813 or so. Bell in the tower dates from 1857.

Kinneigh

GC: 51 45 51 N 8 58 30 W Elev 103 m RR: 60. Ø 5.17 Circumference: 19.23

Unique hexagonal base, local slatestone, dressed to the curve. Door 3.4 high, 9 windows. 20.5 m high. Door 3.24 facing ENE. 4 windows, S, NNE, W, ESE.

Co Donegal

Tory Island

GC: 55 15 51 N 8 13 55 W Elev 4 m RR: 50. Ø 5.17 Circumference: 15.7

Granite beach boulders (undressed). Circumference 15.7 m. Door ESE 2.64 m high. 4 windows, ENE, SSE, WSW and NNW.

Co Down

Drumbo
GC: 54 31 00 N 5 57 35 W Elev 114 m RR: 45. Ø 5.00 Circumference: 15.72
Spalled rubble of local clay slate. 10.25 m high. Door 1.5 m on E. One window on the N.

Maghera
GC: 54 14 17 N 5 53 50 W Elev 6 m RR: 42. Ø 4.85 Circumference: 15.24
Stump of mixed stone, partly granite. 10.25 m high. Door 1.5 m on E. One window on N. OPW made repairs and rebuilding in 1877/78.

Nendrum
GC: 54 29 54 N 5 38 51 W Elev 16 m RR: 30. Ø 4.85 Circumference: 13.3
Stump of mixed stone, partly granite. 4.4 m high. Complete excavation in 1922-4.

Co Dublin

Clondalkin
GC: 54 29 54 N 5 38 51 W Elev 16 m RR: 30. Ø 4.04 Circumference: 12.7.
Complete tower of poorly coursed (not dressed) limestone with some granite. Bulging base. Door 3.9 m facing E. 1 window S 8.18 m. 1 window W 12.57 high. 4 windows, N, S, E, W.

Lusk
GC: 53 31 34 N 6 10 01 W Elev 27 m RR: 40. Ø 5.06 Circumference: 16.0.
Complete to the cornice, 26.56 m, with cap 33 m. Door 0.9 m facing ESE, ground probably raised 2.5m. 6 windows + 2 blocked: NE, ESE, W, NNE, SE.

Swords

GC: 53 27 27 N 6 13 28 W Elev 31 m RR: 50. Ø 5 Circumference: 16.0.

Height 26 meters to the top of the deformed cap. The lintelled doorway, just 70 cm above the present ground level, faces east and is fitted with a modern metal-grilled door. There is a large east-facing window with a slightly jutting sillstone at the second storey level. The 3rd, 4th and 5th floors each have a small square-headed windows to the north, south and west, respectively. At the top storey are four large crumbling windows with flat arches (some with red brick indicating repair work) roughly facing the cardinal compass points. Heavily repaired.

County Fermanagh

Devenish

GC: 54 22 14 N 7 39 21 W Elev 50 m RR: 65. Ø 4.82 Circumference: 15.14.

Excellent condition, well dressed and finely coursed. Sandstone with some limestone internally. Height 25 meters. NE-facing doorway is approximately 2.7 meters above the ground. The arch is composed of three dressed stones running the entire depth of the doorway. A plain raised moulding frames the entire door, including the door sill. The single angle-headed window is slightly to the right and above the doorway at the second storey level. All other windows are lintelled with the third floor window facing NNW, the fourth floor window facing SSE and the fifth floor has the traditional four windows slightly off the cardinal compass points but with a human head carved over each. Repaired in 1835 and 1971.

County Galway

Kilmacduagh

GC: 53 02 59 N 8 53 18 W Elev 18 m RR: 65. Ø 5.68 Circumference: 17.86.

Limestone. Height is just about 34 meters, making it the tallest round tower in existence. The doorway, facing ENE is also extraordinary in that it is almost 8 meters above ground level. Kilmacduagh has 11 angle headed windows, the largest number of windows of any existing round tower. The extreme height of the doorway, the number of bell-storey windows, and the significant lean to the SW all make this a quite unique tower. The cap on the drum is unusual in that it sits not atop a cornice, but overhangs the drum. The walls are over six feet thick at the base. Excavated and repaired in 1878-9, 1971.

County Kerry

Rattoo
GC: 52 26 23 N 9 39 00 W Elev 15 m RR: 70. Ø 4.6 Circumference: 14.5.
Height just over 27 meters. Arched moulded doorway 2.83 meters to the S.E. A single window in the drum of the tower, on the fourth storey - directly above the doorway carved from a single stone. The four large top-storey windows are all angle-headed and roughly face the cardinal compass points. A curvilinear design runs along the top of the arch, resting on the wide raised moulding of the arch itself and terminating in scrolls to either side. The tower is made of a variety of large purple, brown, grey and red sandstone blocks, regularly spaced with smaller stones at intervals. In 1880-81, the OPW reset part of the conical cap. A sheela-na-gig is carved on the top left-hand corner of the north window on the inside.

County Kildare

Castledermot
GC: 52 54 37 N 6 50 05 W Elev 77 m RR: 20. Ø 4.74.
Granite boulders with limestone spalls. Original cap replaced with battlements. Height is 20 meters. Doorway at ground level. The only tower with a window in the same storey as the doorway. There are

two small flat-headed windows in the body of the tower - one to the SE and another to the south. In the top-storey there are four windows, each facing a cardinal compass point.

Kilcullen

GC: 53 06 27 N 6 45 39 W Elev 157 m RR: 35. Ø 5 Circumference: 14.5.

Average height of 10 meters. Local slate stone except for most of the doorway which is granite. Doorway is 1.8 meters above ground level. One rectangular window facing SW.

Kildare

GC: 53 27 27 N 6 13 28 W Elev 31 m RR: 50. Ø 5.3. Circumference: 16.8.

Close on 33 meters high. The doorway faces SSE, a little over 4.5 meters above the ground. The windows in the drum in ascending order face WSW, NW, ESE and SW. The five windows in the storey just below the string course under the battlements face N, NE, ESE, SW and WNW. Conical cap replaced by castellations from the 1730's. Finished and coursed granite in the lowest 3 meters of the tower, contrasting with less evenly coursed and smaller local limestone in the rest of the tower. The doorway is of dark red sandstone carved with chevrons, lozenges and stylized marigolds.

Oughterard

GC: 53 16 39 N 6 33 57 W Elev 138 m RR: 38. Ø 4.58 Circumference: 14.4.

Uncoursed spalled limestone, dressed to the curve. The doorway and arched window are of granite. The tower rises 9.6 meters. The east-facing doorway has a three stone arch devoid of decoration 2.65 meters above ground level. The single window at the second storey level faces south and echoes the design of the doorway.

Taghadoe
GC: 53 21 11 N 7 16 47 W Elev 66 m RR: 40. Ø 4.96 Circumference: 15.6.
Slatey limestone, 19.8 meters high. Doorway SSE, 3.56 m high made of granite. 3 windows, W, SSE and WNW.**County Kilkenny**

Aghaviller
GC: 52 27 54 N 7 16 08 W Elev 85 m RR: 50. Ø 5 Circumference: 15.5.
Slate-coloured sandstone coursed and dressed to the curve inside and out. The almost-level top is 9.6 meters above ground. The original NE-facing doorway is over 4 meters above ground. A single square-headed window facing SW at second storey level.

Fertagh
GC: 52 46 42 N 7 32 41 W Elev 119 m RR: 42. Ø 4.8 Circumference: 15.0.
Height of 31 m. Badly defaced doorway 3.3 meters from the ground NE. It is fitted with a modern opening, possibly by the OPW in 1879-80, having allegedly lost the original stones to a farmer who used them as firebricks in his kitchen in the mistaken belief that they would protect him from fire. There are five windows in the drum of the tower. Two are angle-headed and three are lintelled, SSE, N, W, NE, SSE. The four angle-headed windows in the top-storey face the cardinal compass points

Kilkenny
GC: 52 39 23 N 7 15 25 W Elev 58 m RR: 58. Ø 4.5 Circumference: 14.0.
Limestone. An excavation of 1846-47 found that the foundations extend less than three feet down. Height of just over 30 meters. The SSE facing arched doorway with no decoration and 2.7 meters above ground. The windows, all lintelled and in ascending order, face NW, NE, just to the E of S and just to the S of W. The top storey

below the parapet has six evenly spaced windows, unique, except for the round tower at Kilmacduagh.

Kilree

GC: 53 27 27 N 6 13 28 W Elev 31 m RR: 50. Ø 4.86 Circumference: 15.28.

Height is between 26 and 27 meters. The arched doorway is 1.64 meters above the ground and faces south. Limestone, sandstone doorway. Two windows in the drum are to the north and east, both lintelled. The top-storey has four lintelled windows facing the compass points.

Tullaherin

GC: 52 34 45 N 7 07 48 W Elev 83 m RR: 46. Ø 4.92 Circumference: 15.5.

Height is 22.5 meters from ground. Dressed breccia. Windows are all lintelled and face, in ascending order, SE, W, WNW, and NNE. The top-storey, which appears to have had eight windows similar to O'Rourke's tower at Clonmacnoise has four remaining windows, all rectangular and lintelled. They face NW, N, NE and E. The doorway with no casing supported by a stabilizing pier inserted in 1892, 3.7 meters above ground. A capless tower, parapetted.

County Laois

Timahoe

GC: 52 57 37 N 7 12 13 W Elev 121 m RR: 43. Ø 5.5 Circumference: 17.5.

Height is 29.26 meters. The receding doorway is 4.9 meters above ground level facing ENE. The lowest window, facing S is also unusually finished, being large and framed in stone. The window mid-way on the rear of the drum is a lintelled slit and the top drum window is squared and lintelled, facing WNW. The four top-storey windows face the cardinal compass points. Renovated by the OPW in the late 19th century when the cap was repaired and a modern ground level doorway was filled on the SW side of the tower. A complete tower without floors or ladders. It has one of the finest four-order Romanesque doorways in Ireland, with elaborately carved and decorated with interlace, human heads, chevrons and capitals. It is unique in round tower architecture.

County Limerick

Ardpatrick
GC: 52 20 18 N 8 31 56 W Elev 218 m RR: 25. Circumference: 16.5.
A stump with dressed and squared stones.

Dysert Oenghusa
GC: 52 31 15 N 8 44 41 W Elev 22 m RR: 42. Ø 5.28 Circumference: 16.6.
Limestone with door and windows of sandstone. OPW work in 1881-2. Door facing E 4.6 meters above ground. 3 windows, W, SSW, NNE.

Kilmallock
GC: 52 24 04 N 8 34 29 W Elev 86 m RR: 35. Ø 5.28 Circumference: 16.6.
Incorporated into the church.

County Louth

Dromiskin
GC: 53 55 19 N 6 23 53 W Elev 13 m RR: 38. Ø 5.22 Circumference: 16.4.
Slatestone. Height 15.24. OPW work in 1879-80 to the already squat tower. Door 3.7 ESE. 1 window WNW. 4 recent windows at the top.

Monasterboice
GC: 53 55 19 N 6 23 53 W Elev 13 m RR: 38. Ø 4.98 Circumference: 15.63.

Slatestone. Height 28 meters. The east-facing arched doorway is just 1.84 m. above ground level. A wide double banded moulding frames the doorway. Aside from the beautifully crafted quoins, there is no other decoration. Like the doorway, the angle-headed window directly above it is composed of sandstone. The three other windows in the drum are lintelled and small, facing - in ascending order - W, S, and N. Apparently is still equipped with floors and ladders, but while there is a concrete and metal stairway to the door, the iron grille fitted into the doorway is padlocked.

County Mayo

Aghagower
GC: 53 55 19 N 6 23 53 W Elev 13 m RR: 38. Ø 5.00 Circumference: 15.76.
Limestone. Height 15.85. Rebuilt in 1969. Doorway 2.18 above ground faces
east. There are three windows, SSW, WSW and S. No roof. The modern ground floor doorway allows entry into the tower.

Balla
GC: 53 48 18 N 9 7 53 W Elev 38 m RR: 27. Ø 5.26 Circumference: 16.56.
Sandstone. Height 10 m. Two doorways in this abbreviated tower. The upper, and possibly original doorway is placed higher than almost any other round tower, almost 8 meters above the present ground level. There is one tiny arched window on the south side of the tower, perfectly carved using four stones, about halfway between what would have been the first and second floors of the tower.

Killala
GC: 54 12 56 N 9 13 14 W Elev 10 m RR: 55. Ø 5.02 Circumference: 15.78.

Limestone with some very large stones. Height 25.6. Round doorway SSE 3.8 meters up. There are three square windows in the drum face (in ascending order) ENE, SSE and W. The traditional four windows in the top storey are square-headed internally and angle-headed externally and do not face the cardinal compass points, but are slightly skewed, facing NNE, ESE, SSW, and WNW. The OPW did some repair work in 1841, the cap and wall were repaired at this time.

Meelick

GC: 53 55 17 N 9 01 12 W Elev 34 m RR: 37. Ø 5.45 Circumference: 17.1.

Quartzy sandstone with some limestone. Height 21.3. The five stone arched doorway is 3.42 meters and faces SSE. Of the six windows, the lowest two are angle-headed facing ENE and S, and the four remaining lintelled windows face NNW, WSW, ESE and N. No cap or top storey. OPW did some work repointing in 1880-81, but details are not recorded.

Turlough

GC: 53 53 18 N 9 12 30 W Elev 34 m RR: 50. Ø 5.5 Circumference: 17.5.

Quartzy sandstone with some limestone. Height 22.86 m. The arched doorway (currently blocked with mortared stone) is 3.96 m above the ground level below it to the SE. Angle-headed top-storey windows face just to the left of the cardinal compass points. The other small lintelled windows in the drum from bottom to top are oriented to the south, then slightly skewed to the west, north and east. It was repaired in 1880 by the OPW.

County Meath

Donaghmore

GC: 53 40 13 N 6 39 44 W Elev 51 m RR: 68. Ø 4.98 Circumference: 15.65.

Limestone. Height 24.7 meters. The rounded, sandstone east- facing doorway is about 3.4 meters above ground, contested crucifixion carving Petrie for, Windele against. There are four windows which, in order, are a south-facing lintelled window, angle-headed east window, an arched west window and another lintelled window just below what would be considered the top-storey. The tower was reportedly repaired by an owner in 1841 and some restoration work also done by the OPW at a later date.

Kells
GC: 53 43 38 N 6 52 48 W Elev 84 m RR: 55. Ø 4.95 or so Circumference: 15.5.

Limestone. Height 26 meters. Round-headed doorway is 3.3 above the offset and 1.86 meters the other side, facing N. There are four small lintelled windows in the drum, facing (from bottom to top) SSW, ESE, N, and WNW. The top storey windows of the tower are unusual in that there are five of them, rather than the usual four, Kildare being the exception, evenly spaced, they face ENE, SE, SSW, W and NNW.

County Monaghan

Clones
GC: 54 10 39 N 7 13 58 W Elev 58 m RR: 42. Ø 4.98 Circumference: 15.39.

Sandstone. Height 15.4 m depending on the base. The east-facing doorway is 1.64 meters above the present level and 2.12 m above the offset. Small lintelled windows face - in ascending order - S, N and E with the traditional but probably late four top-storey windows at the cardinal compass points.

County Offaly

Clonmacnoise
GC: 53 19 35 N 7 59 11 W Elev 39 m RR: 68. Ø 5.62 Circumference: 17.66.

Limestone. Height 19.3 meters. The doorway faces southeast at a height of 3.5 meters above the offset at the base. It is a slightly triangular shaped arch, being narrower at the top, wider at the bottom. The arch is beautifully cut of nine matching stones resting on transitional stones that project slightly into the doorway and above the five jambstones on either side. All stones carry through the entire depth of the wall. There are 10 windows. One lintelled window to the northwest is one level above the doorway. Another lintelled window is to the north facing the River Shannon. The other eight windows, all lintelled, facing the cardinal compass points and opposing points between.

County Sligo

Drumcliff
GC: 54 12 56 N 9 13 14 W Elev 10 m RR: 55. Ø 4.98 Circumference: 15.65.

Height 8.8 m. The ESE-facing doorway is 1.75 meters above ground level. One very small lintelled window survives facing SSW at about the second floor level.

County Tipperary

Cashel
GC: 52 31 13 N 7 53 25 W Elev 122 m RR: 57. Ø 5.3 Circumference: 16.75.

Sandstone. Height 27.94. Excavations done in 1841 found the tower built directly on the solid rock outcropping. The original doorway faces SE, 3.28 m above ground. The arch is composed of seven stones with five stones in the west jamb and six in the east. Three square-headed, lintelled windows in the body of the tower and four angle-headed windows in the top-storey, off the cardinal compass points in the NE, NW, SE and SW. Two of these angle-headed windows are formed with a single stone, unlike the more common configuration of two stones set to either side forming a 90 degree angle.

Roscrea

GC: 52 57 21 N 7 47 45 W Elev 96 m RR: 68. Ø 4.6 Circumference: 15.25.

Limestone. Height 20m. No cap. The arched, SSE-facing doorway is just about three meters above the present ground level. The large east-facing window in what would have been the second storey is angle-headed. It is notable for the single-masted ship carved in relief on the north jamb toward the inner edge. There are two other windows in the drum of the tower. Both are much smaller and lintelled with the lower of these facing west and the upper, north.

County Waterford

Ardmore

GC: 51 56 54 N 7 43 34 W Elev 37 m RR: 72. Ø 5 Circumference: 15.8.

Sandstone. Height 26 m. Door ENE 3.78 m above ground. Seven windows, facing N, ENE and SSW, four top-floor windows on cardinal points. Unique feature of three string courses, which are rounded and project in an unbroken circle approx. 6 meters from ground level for the first course, 11 m above ground level for the 2nd course, and about 17.5 m above ground level for the third course. Unique carved corbels, five of the 16.

County Wicklow

Glandalough

GC: 53 00 38 N 6 19 40 W Elev 138 m RR: 72. Ø 4.87 Circumference: 15.3.

Mica schist and some granite. Complete with new (1876) cap. 30.48 high. Door SSE 3.2 above the offset. 4 windows, SW, NW, NNE and ESE. 4 larger windows at the top, NNE, ESE, SSW and WNW.

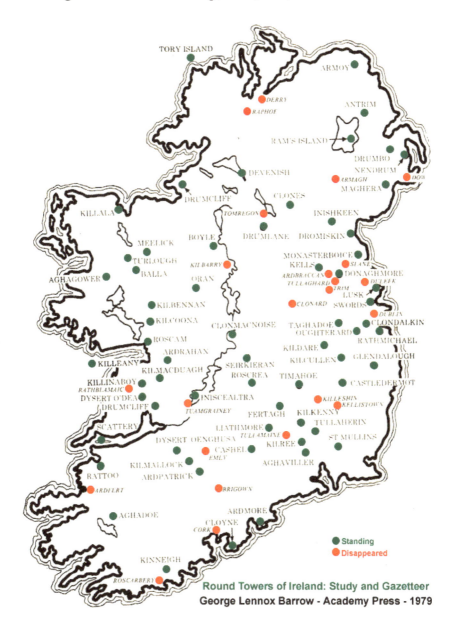

Round Towers of Ireland: Study and Gazetteer
George Lennox Barrow - Academy Press - 1979

Scotland

Abernethy
GC: 56 19 58 N 3 18 42 W Elev 37 m RR: 37. Ø 4.66 Circumference: 14.63.
Sandstone. Height 22 m. N-facing door. 3 windows, S, W and E, and 4 larger windows at the top on cardinal points.

Brechin
GC: 56 43 50 N 2 39 41 W Elev 53 m RR: 37. Ø 4.54 Circumference: 14.3.
Sandstone. Height 25.96. Arched door facing W, 2.19 above ground, with carvings. 2 windows E and S in the barrel and four to cardinal points at the top, with later additions above. Elaborate carved door with a tau cross.

There seem to be 23 towers which have been demolished or destroyed, at least no longer there, but reference has been made to them at some stage in the past. There seems little point in giving the locality or other detail, for such can be found in Barrow's or Lalor's books.

On the basis of the principles to be found in authentic towers, as expressed in this essay, whereby the stones are systematically laid with the negative magnetic polarity facing upwards and over the crossing of underground water, some towers are eliminated from the list generally accepted, this would include the towers at Ferns, on Ram's Island, Egilsay in the Orkneys and St Patrick's Isle, Peel, Isle of Man.

Appendix III

Cyclopean or Pelasgic masonry

Three photos to show how the Polarity Principle is used in different structures around the world. Negative side up! And some examples of the Cyclopean interlocking stonework which ensures solidity of the structure.

Mycenae Gate, Greece

Machu Picchu, Inca Palace

Sacsahuamán (Sacsayhuamán), Cusco, Peru. The Tiopunco
portal, one of the three portals

Appendix IV

Readings from experiments with stones in Chiang Mai

Experiments conducted in Mae Rim on 20 and 21 November 2009 on a piece of clear land in the shade of tamarind trees. The average temperature was approximately 26 degrees C and an atmospheric pressure of 1,900 or 28.8. The reading of the location was 6,500 on the Bovis scale and colour vibration green. The rocks used were taken from Wat U-Mong, probably from a local river and the pendulum says they are in the region of 450,000 years old.

Square formation with 4 right angles
4 stones placed at an equal distance of 1.618 m from each other and + facing up.

A – 15,000   B – 15,500   A-B facing North, B-C facing East, C-D facing South
C – 15,500   D – 15,000   D-A facing West
Readings:
Centre 26,000   Colour: Green

Square formation with 4 right angles
4 stones placed at an equal distance of 1.618 m from each other and + facing up.

A – 15,000   B – 15,500   A-B not facing North
C – 15,500   D – 15,000
Readings:

Centre 26,000   Colour: Green
40 m from centre 26,000 in all directions. 50 m from centre 25,000 in all directions. 70 m East 22,000, 80m East 20,000, 90 m East 16,000 and 120m East 8,000

Square formation with 4 right angles

4 stones placed at an equal distance of 1.618 m from each other and + facing up.

A – 15,000   B – 15,500   A-B not facing North and A inversed
C – 15,500   D – 15,000
Readings:
Centre 20,000   Colour: Blue
5m South 16,000

Square formation with 4 right angles

4 stones placed at an equal distance of 0.99 m from each other and + facing up.

A – 15,000   B – 15,500   A-B facing North
C – 15,500   D – 15,000
Readings:
Centre 15,000   Colour: Blue
50m South 11,500

Square formation with 4 right angles

4 stones placed at an equal distance of 0.99 m from each other and + facing up.

A – 15,000   B – 15,500   A-B facing North and A inversed
C – 15,500   D – 15,000
Readings:
Centre 9,000

Equilateral triangular formation

A – 15,000, B – 15,500 and C – 15,500 placed at 1 m distance
Readings:

Centre 27,500   Colour: Green
50m South 27,000
120 m East 15,000

Isosceles triangular formation
A – 15,000, B – 15,500 and C – 15,500 A placed perpendicular
0.618m from B-D at 1 m distance
Readings:
Centre 31,000   Colour: Green

Circle formation with a radius of 1.618 m using 13 stones of an average 10,000 B.U. + up
Readings
Centre 34,000   Colour: Green

Circle formation with a radius of 1.618 m using 13 stones of an average 10,000 B.U. + up and 1 stone of 15,000 B.U. in the centre + up
Readings:
Centre 45,000   Colour: White
 from centre 45,000 in all directions. 120m East 18,000

Circle formation with a radius of 1.618 m using 13 stones of an average 10,000 B.U. + up and 1 stone of 15,000 B.U. in the centre - up
Readings:
Centre 25,000   Colour: White

 Circle formation with a radius of 1.618 m using 12 stones of an average 10,000 B.U. + up and 1 stone of 15,000 B.U. in the centre + up
4 stones placed at an equal distance of 0.99 m from each other and + facing up.
Readings:
Centre 51,000   Colour: White
50m South 51,000 colour white. 120 m 27,000

Circle formation with a radius of 1.618 m using 13 stones of an average 10,000 B.U. + up

Readings:

Centre 31,000   Colour: Green

Circle formation with a radius of 1.618 m using 18 stones of an average 10,000 B.U. + up and 3 stones of 15,000 B.U. in the centre + up

Readings:

Centre 90,000   Colour: White

50m South & East, 90,000 colour white. 120 m 45,000, 150 m 30,000.

Circle formation with a radius of 1.618 m using 18 stones of an average 10,000 B.U. + up and 2 stones of 15,000 B.U. in the centre + up and 1 stone of 15,000 B.U. - up

Readings:

Centre 45,000   Colour: White

A "hot-spot" 44 m from the experimenting area, normally reading 2,000 B.U. rose to 22,000 B.U. under the influence of the square 4-stone (1.618 m) influence.

When placing the stones with the – polarity facing up, the increase in the score of the average reading was of 50%.

Appendix V

Test of the rock magnetic polarity

Negative magnetic polarity facing upwards.

Score of 70 on my chart.

APPENDIX VI

Kilmacduagh, Co Galway. The doorway is at 7.92 m from the offset.

APPENDIX VII

RADIESTHESIA

How does a human obtain the answer to a question? That perhaps appears to be an incredibly silly question, but have you ever considered the process? No matter whether intuition or belief in authority is your go-to source, there is a strong chance that you have, perhaps, never even paid attention to this all-important topic.

I would venture that people consciously avoid such hard questions, we are educated, indoctrinated and encouraged to accept what we are told, and of course we know that it is the role of the unruly, deranging adolescent to make a speciality of questioning authority. As a consequence, we soon learn to develop arguments and a so-called rationale to make sure we remain ensconced in our comfort zone where nothing is disturbed, or can disturb, the sacred *status quo*, and no further research is made into epistemology, the name we give to that beast.

It is most uncomfortable to not know. The whole range of human emotion can and does find expression in that space – from the self-assumed humiliation of not providing the right answer to a simple question in class, to the lifelong motivation generated by the desire to know what it is that sustains life; so potentially an important area to have well covered by whatever means available. But those means are tenuous and entirely subjective, most often based on trust, although not much thought goes into the process. It is hardly surprising that we have reached the current state of society where opinion has become the sole human virtue. However, a great many people are not bothered by the discomfort of living in the untruth, no thinking, no responsibility, no ethical values to defend, let's go have another beer!

But what if a phenomenal, sure-fire method to access truthful answers to questions exists?

I believe there is, and it has been around for ever. While it is apparently easy enough to access, the discipline needed is not to the liking of most. Three hours a day for two years – every day, I would add – is what it took me to achieve the confidence that seemed essential to rely 100% on the method. I still make mistakes, and generally they are due to a poor phrasing of the question, perhaps its motivation, and such like.

A method that can lead to that unique peace of mind that comes with being in the truth, even if it does involve a period of residence in heresy – the non-acceptance of mainstream "science", which challenges the simplest precepts and notions on which modernity – and tradition for that matter are based, is not going to earn you too many friends, and perhaps some enemies, so very likely a lonely exercise.

There is no need here for anything of a philosophical, psychological, or psychic nature. There is no requirement for a belief system. It is a totally practical, simple method that requires, as mentioned above, a vast amount of practice; combined with careful thought in formulating the questions; discipline and – at some stage – a quiet conviction as to how the whole process works. This last component is of the essence, the consequence of truth becoming apparent repeatedly via answers to multiple questions that have, in time and context, been revealed in their true verifiable nature. Now there is a tool worth using!

If you are sufficiently familiar with radiesthesia, skip this section; if not, please read this passage mainly extracted from *Radiesthesia I,* my pamphlet concerning how it might work.

RADIESTHESIA OR HOW TO DEVELOP WHOLISTIC AWARENESS

Radiesthesia is much more than a method for finding answers to questions.

An image springs to mind perhaps - the guy with the hazel twig looking for water; the soldier with the rods looking for tunnels; or yet again, a person sitting at their desk, pendulum in hand, bent over a map searching for mineral deposits in some distant land. Those are not just possibilities. They are actual. Much like the astrologer working in a bank, deciding the right moment to invest, in what and how much.

No matter what your objective, if you are seriously considering adding this discipline to your toolbox, and obtaining the right answers to your questions, it would be best that, right from the start, you approach the process as a discipline. There are no two ways about it, if you hope to achieve the confidence that makes the difference between a good and indifferent radiesthesist, you need not only an awful lot of practice but a constant awareness of communion – communion with the essence. This is a divining art. We know, in all humility, that we are in a state of total ignorance, which must presumably be the case if we do not 'know' if the apple is nutritionally good for us, or just good taste-wise, otherwise why would you ask? So, why not approach the divine for inspiration, if not the answer?

You are working with what really exists, not with the mental constructs which we often take for our reality.

It is a complex process, when viewed from the viewpoint of mental implication; it is even more complex when viewed from the physical standpoint. That is perfectly normal, because we do not know very much about anything, which is perfectly normal, we are part of it.

Most people who use a pendulum share the belief that everything that exists has its specific frequency or vibration. Now if you relate to that frequency, and why not, you are in a direct link with that object, and by definition no longer "connected" to whatever else is in the vicinity. This frequential approach is known as the physicist school of radiesthesia, very closely related to modern science, where everything is dissociated and inspected in its own light, so to speak.

The mentalist school – the other school defined by the French, works on a much more psychological principle, whereby the intellectual faculty of the mind asks a question of the sub-conscious component of the mind because it does not know the answer, and thanks to a neuro-muscular response, a reply comes from some undefined entity which in theory is in communication with the 'sub-conscious'.

The antagonism between the two is easy to understand, and is perfectly mutual.

If, on the contrary, you consider yourself, as I do, to be a part of, rather than apart from, everything that exists, the process becomes remarkably different. As of the time that you recognize that you don't know, you are potentially open to whatever is actually there. But if you focus on a specific frequency, you again create a subject-object relationship. That is not where one needs to be. It seems reasonable to say that the best possible position to receive quality information is in a state of vacuity, with nothing imposed on the mind space environment. There is a greater chance of connecting with the total environment (rather than the mental space and whatever is there) when you feel you are a component rather than some sort of controlling influential factor. The danger, however, is a lack of vigilance because that vacuity can be invested by uninvited ideas or forms. Consequently, a strong sense of discrimination is essential, to be employed at all times to ensure its firm anchoring in the state of being of the radiesthesist, whereby awareness or recognition of any impression coming into the mind-space is calmly acknowledged. Such a state can be described as being open to an impression, the sense of radiesthesia, as given by the two French churchmen, Boulé and Bayard in the early 20th century.

Neither the mentalist nor physicist schools – as defined by the French in the early 20th century – fit this open mindset. The two are mutually exclusive, and both fall short of the total reception of the required perceiving. What is worse, to my mind, it does not correspond to a wholistic approach where inclusion is the key. Especially inclusion of one's thought process or mind, as one of many similar components that make up the human segment of this dimension. Which although highly important to us, is of little consequence to the flow of events from a cosmic viewpoint!

At some stage, and it will probably be early on, once you have acquired a certain ease in using a pendulum, you will ask *"What's going on here?"* I make no claims to knowing any more than the next person. In fact, on the contrary, I am such an ignorant fellow that I have to use a pendulum to discover answers to the simplest of questions!

The passage of time has taught me one thing.

Things change. Constantly.

But we do have this remarkable means of discovering the truth of the moment. Would it not be foolhardy not to access a possible answer if it is so readily available?

An explanation as to how a neuro-muscular response to a mentally formulated question can occur would be welcome. Or, at least a convincing theory as to its apparent working.

There is one thing common to life – all life. It is consistently subject to natural forces. No matter how hard one tries to work with or against those energies, they resist. They have a life force of their own, in the same way that we do. One might even say they have an intelligence. Indeed, they do, but that is not what science would have us believe, for if anyone who has studied and worked with water will tell you, that is the only honest conclusion one can reach. When one stops being a person, a scientist, a dowser, a compartmentalized individual, there arises a chance – I repeat, a chance – that one is. "Is" as in "I am". In a materialistic regime, for the rest of our time here on earth, that angle is reinforced but rarely examined. In a more spiritual, wholistic regime, that is the angle to be examined and pared down to size.

The natural conclusion is a complete unity, a perfect totality with everything in its place, in constant but impeccable flux. Thank goodness, we do not see this as it is, it would be far too frightening for our fragile constituencies.

The whole point, however, is that there is a component which we perceive and feel, and another which we merely feel. The conscious and the (rather rudely termed) unconscious.

Radiesthesia is the bridge between these two: the phenomenally aware, conscious component, which might be equated with the cerebral mind, whereas the real brain is probably the entire human metabolism; and the much broader, because englobing all components of our physical reality, all-conscious (as I prefer to call it).

Of course, there is no differentiation, therefore no bridge, but our human ego separates you from me, and we spend a life-time trying to establish that bridge between two continents that were never apart in the first place!

If one can accept such a theory, there is a substantial advantage in that the phenomenon of dowsing does not have to depend on some elaborate explanation of physics (based on differentiation) or magic, which although it always will be, you no longer have to rely on some undefined, unknown – therefore mysterious, factor.

It is this mysterious side which gets any 'unscientific' phenomenon a bad name. If you cannot verify something, it is very easy to make a claim and attribute it to some peculiar property of which you are, of course, the sole guardian.

This approach to radiesthesia, based on communication between the conscious and the all-conscious, removes all need for any mystification, and at the same time puts mystification exactly where it should be. In the realm of egocentricity and real pseudo-science!

There might well be something of all this to be found in ancient cave drawings. Whether from Australia, China, France, Patagonia, South Africa or Spain, the paintings often portray a connection established between the subject - humans in the case of cave art (human hand-prints, hunters with bows, spears and assorted weapons), and the object – animals, or more specifically – lunch.

As with so many aspects of life, some people are gifted. Probably in those days – forty thousand years or more ago, the same held true. Some men or women were better able to locate the group of animals to be found – or avoided. Those individuals were no doubt encouraged to hone their capacity and were surely very useful members of society. Is that perhaps how they consolidated their skill, on the walls of their homes?

The connection enabling a link between an unseen source of life and our frequently urgent need to access the information as to the whereabouts of that energy source, is as real today as it was in the past. The means of tapping into it have changed, and in the process, the natural ability of "primitive" humans has declined. As often happens when a sense is not used, it atrophies and access is obscured.

That does not mean the information is no longer accessible. On the contrary, I believe, it is and always will be. But this 'sense' needs to be acknowledged, nurtured and developed.

Though it may sound trite to say so, we humans are equipped with a logical reasoning capacity. That is why we know when we don't know something. We have an ability to learn from our mistakes – well, some of us anyway. It is my firm conviction that we, as humans, in this modern age, do not question sufficiently. Not only do we not question whether something is right or wrong, but we do not doubt what an authoritative voice tells us. That is dangerous and tragic. It is as if we have lost the ability to discriminate what is good for us, or not.

In a manner of speaking, that is what has happened. We no longer have the know-how to obtain an honest answer to a simple question. The brain rules and we are led by the nose along the path of our existence that can only lead to our undoing if we fail to learn that it is the heart, our intuitive connection, which precedes. Would it not be more sensible to follow that path in the knowledge that it is the right one for us – individually or collectively – thanks to our capacity to discern?

The pendulum is surely the most flexible of all the tools that can be used to find answers to questions. It is *par excellence* the "barefoot" way to discover if something (food, drink, person, medication, situation) is beneficial or not for you or others. Because of its flexibility and simplicity of use, the pendulum rather than other methods, such as dowsing rods, the hazel twig or equivalent, or again the Bi-digital O-ring, friction, arm muscle, sway, toe-touching tests, although equally valid, shall not be considered here.

––––––––––––––

There is very little that cannot be determined using a pendulum, and not a single domain in our physical life where it cannot be applied, though becoming confident in its use requires a lot of work; involving not just practice, but understanding, if one is to acquire the desired intuition. However, almost everyone can become proficient in its use.

As Malcolm Rae, the 'father' of modern radionics, also a radiesthesist, said:

> "One of the most alarming features of Radiesthesia is the number of people who, finding some sort of response from a pendulum, believe themselves competent to use one, completely and accurately, whilst possessing nothing approaching the skill and reliability required for responsible use. There is something in common between learning to use the pendulum and learning to play the piano, in that skill comes only from an adequate amount of practise in doing each well.

> "No-one believes that the ability to press a few notes on the keyboard of a piano and thus to produce a few simple notes sometimes recognisably similar to those in a known piece of music indicates an ability to play complicated music in a faultless manner. No-one should entertain a similar misbelief about a pendulum - for in both cases practice is necessary. Possibly the main difference is that the inaccurate pianist can do much less harm than the inaccurate Radiesthesist for errors of the former are much more obvious."

There are a number of human qualities that would be well to acquire beforehand. Firstly, because the answers can only come one at a time, and only in the form of yes or no. Be kind on yourself, take your time! The overall picture will be completed with patient questioning but remember, the process remains binary at all stages.

APPENDIX VIII

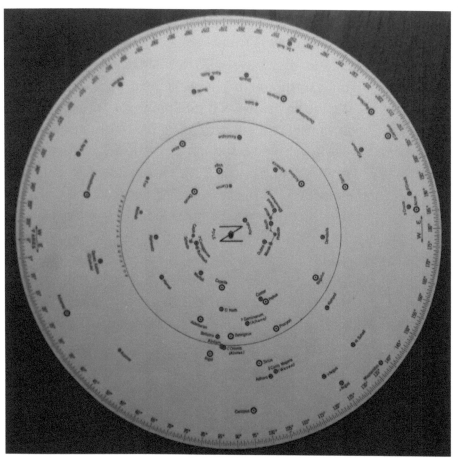

Photo of the northern hemisphere chart from Rude's Star Finder and Identifier, published at the Hydrographic Office, Washington DC in March 1942.

ABOUT THE AUTHOR

Numerous avatars, some divine, others more mundane, across the world saw the man transform from an officer in the British Gurkhas in the late sixties, and after travelling overland to Zambia where he worked as a guide on photographic safaris on foot and met his preceptor, he became an orthodox Hindu monk for four years. That formed the basis for his real education leading to in-depth studies of Vedanta, Sanskrit, ayurveda, and Indian culture. Fast forward to France, after periods spent in the USA, Europe, Asia and the Far East, he settles as a technical French-English translator, the compiler of three dictionaries, before globalization throws the whole translation world into disarray, and he returns to the study of geomancy, radiesthesia and Chinese medicine. An eleven-year stint in Thailand results in his becoming a full-time radiesthesist actively applying the lessons learned along the way to resolving geopathic stress, providing solutions for health, researching concerns both natural and man-made, and such related matters. Now living in the depths of Ireland running workshops, writing and offering his devices and services to those in need.

Previous self-publications include:

- *Other-Dimensional Entities,* the recognition and release of spirit attachments, 2022.
- *The Way of the Skeptic,* a wholistic approach to some of the key factors involved in living, 2017.
- *Embracing the Sublime,* an autobiography, 2019.

- *Radiesthesia I,* method & training for the modern dowser, 2019.
- *Radiesthesia II,* medical dowsing and the vital force in the body, 2019.
- *Radiesthesia III,* an educated guess at the 365 senses available to humans, 2020.
- *Radiesthesia IV,* geopathic stress – insights, forms and solutions, 2020.
- *The Last Word*, with Swami Pranav Tirtha, a translation of the *Yoga Vasishta Ramayana*, an 11th century Sanskrit text on the essential teachings of Vedanta, 1975.
- A *Sanskrit-English Philosophical Wordlist,* a dictionary of terminology (3,000 terms) specific to Hindu and related philosophies, cosmology and mythology, 1977.
- *Dictionnaire Pratique des Mondes de la Finance et de la Bourse,* a French-English, English-French lexicon of stock market and financial terminology of over 10,000 terms, 1996.
- *Vade-mecum Bilingue des Exploitants de Centrales Nucléaires,* with Jean Miron, a robotics engineer at EDF. Unpublished to date.

CPSIA information can be obtained
at www.ICGtesting.com
Printed in the USA
LVHW070021290323
742731LV00050B/258